Yves André

G-Functions and Geometry

Aspects of Mathematics
Aspekte der Mathematik

Editor: Klas Diederich

All volumes of the series are listed on pages 230–231.

Yves André

G-Functions and Geometry

A Publication of the Max-Planck-Institut für Mathematik, Bonn
Adviser: Friedrich Hirzebruch

V

Springer Fachmedien Wiesbaden GmbH

Dr. *Yves André*

Institut H. Poincaré,
UA 763 du CNRS
11 rue P. et M. Curie
75231 Paris 5

AMS Subject classification: 11 G xx, 11 J xx

Originally published by Friedr. Vieweg & Sohn Verlagsgesellschaft mbH, Braunschweig in 1989.

Produced by Wilhelm + Adam, Heusenstamm

ISSN 0179-2156
ISBN 978-3-528-06317-7 ISBN 978-3-663-14108-2 (eBook)
DOI 10.1007/978-3-663-14108-2

Foreword

This is an introduction to some geometric aspects of G-function theory. Most of the results presented here appear in print for the first time; hence this text is something intermediate between a standard monograph and a research article; it is not a complete survey of the topic.

Except for geometric chapters (I.3.3, II, IX, X), I have tried to keep it reasonably self-contained; for instance, the second part may be used as an introduction to p-adic analysis, starting from a few basic facts which are recalled in IV.1.1. I have included about forty exercises, most of them giving some complements to the main text.

Acknowledgements

This book was written during a stay at the *Max-Planck-Institut* in Bonn. I should like here to express my special gratitude to this institute and its director, *F. Hirzebruch*, for their generous hospitality. *G. Wüstholz* has suggested the whole project and made its realization possible, and this book would not exist without his help; I thank him heartily. I also thank *D. Bertrand, E. Bombieri, K. Diederich,* and *S. Lang* for their encouragements, and *D. Bertrand, G. Christol* and *H. Esnault* for stimulating conversations and their help in removing some inaccuracies after a careful reading of parts of the text (any remaining error is however my sole responsibility).

It is a pleasure to acknowledge the influence of previous work of *Bombieri; Christol,* and *G. V. Chudnovsky* on this book. Finally, I wish to thank Miss *Grau* for her patience in deciphering and typing the whole manuscript.

May 1988 *Yves André*

Contents

Logical Dependence of the Chapters

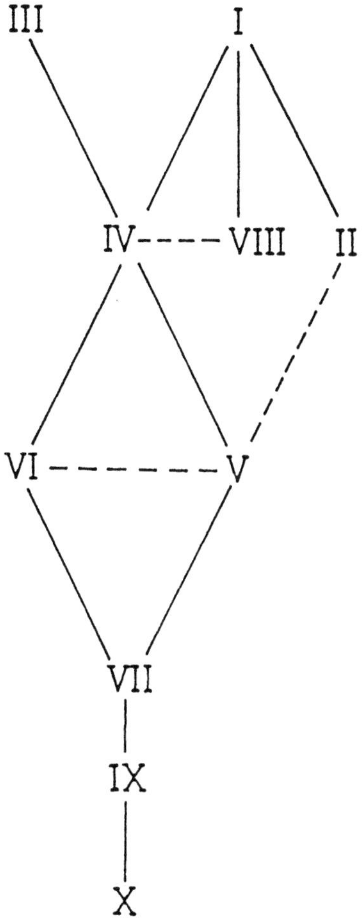

Notations

GENERAL NOTATIONS. $\mathbb{N}$ is the set of natural numbers; $\mathbb{Z}$ (resp. $\mathbb{Q}$, $\mathbb{R}$, $\mathbb{C}$) is the ring)resp. the field) of integers (resp. of rational numbers, of real numbers, of complex numbers). If p is a prime number, $\mathbb{F}_p$ denotes the prime field $\mathbb{Z}/_{p\mathbb{Z}}$ and $\mathbb{Z}_p$ (resp. $\mathbb{Q}_p$) the ring of p-adic integers (resp. the field of p-adic rational numbers). For $t \in \mathbb{R}$, we shall write $\log^+ t$ for $\log \mathrm{Max}(1,t)$; one has $\log^+ t_1 t_2 \leqq \log^+ t_1 + \log^+ t_2$. We denote by $[t]$ the integral part of t : $[t] \in \mathbb{Z}$, $[t] \leqq t < [t] + 1$. We denote by $\overline{\lim}$ (resp. $\underline{\lim}$) the upper (resp. lower) limit of a sequence of real numbers. If f,g are two functions of a real variable, with $g \geqq 0$, we write $f = O(g)$ if there exists a constant $C > 0$ such that $|f(x)| \leqq Cg(x)$ for all sufficiently large x ; we write $f = o(g)$ (resp. $f \sim g$) if $\lim_{x\to\infty} f(x)/_{g(x)} = 0$ (resp. 1).

PLACES

Symbols:

$\overline{\mathbb{Q}}$	a fixed algebraic closure of the field of rational numbers,
K	a number field; that is to say, a subfield of $\overline{\mathbb{Q}}$ which is a finite extension of $\mathbb{Q}$,
$\mathcal{O}_K$	the ring of integers in K,
$d = [K:\mathbb{Q}]$	the degree of K over $\mathbb{Q}$,
Σ or $\Sigma(K)$	the set of all places of K,
Σ_f (resp. Σ_∞)	the subset of finite (resp. infinite) places,
$v\|p$ or $p = p(v)$	v lies above the place p of $\mathbb{Q}$,
K_v	a completion of K with respect to $v \in \Sigma$,

$d_v = [K_v:\mathbb{Q}_{p(v)}]$ — the local degree at $v \in \Sigma$; one has $d = \sum_{v|p} d_v$.

Normalization:

$| \ |_v$ — the absolute value in K_v normalized in the following way:

$$|p(v)|_v = p(v)^{-d_v/d} \quad \text{if } v \in \Sigma_f \text{ (ultrametric case)},$$

$$|\xi|_v = |\xi|^{d_v/d} \quad \text{if } v \in \Sigma_\infty \text{ (Archimedean case), where}$$

$| \ |$ — denotes the Euclidean absolute value on K_v , for $v \in \Sigma_\infty$,

$\mathbb{C}_v$ — a completion of an algebraic closure of K_v ; $| \ |_v$ extends to $\mathbb{C}_v$,

$i_v : K \hookrightarrow \mathbb{C}_v$ or K_v — the natural imbedding.

Remarks:

The symbol $\sum_v$ will denote a summation with all $v \in \Sigma(K)$. For any finite extension K' of K , any $\zeta \in K$ and $v \in \Sigma(K)$, one has $|\zeta|_v = \prod_{w \in \Sigma(K')} |\zeta|_{K',w}$, and all factors have the same value; see [43],[54] for this material.

RINGS. Let R be a commutative entire ring with unit. We shall use the following entire rings (with standard operations):

$R[x]$ — the polynomial ring over R ; more generally,

$R[\underline{x}]$ — the polynomial ring in several commuting indeterminates $\underline{x} = (x_1,\ldots,x_\nu)$ over R ,

$R(x)$ — the fraction field of $R[x]$,

$R[[x]]$ — the ring of formal powers series over R ,

$R((x))$ — the fraction field of $R[[x]]$,

$M_\mu(R)$	the ring of square matrices of size μ over R ; we shall identify $M_\mu(R((x)))$ with $M_\mu(R)((x))$,
$GL_\mu(R)$	the group of its invertible elements,
I or I_μ	its unit,
$\binom{Y}{n}$	for $Y \in M_\mu(R)$, $\binom{Y}{n} = (n!)^{-1}Y(Y-I)\dots(Y-(n-1)I)$ whenever $n!$ is invertible in R .
${}^tY$	the transposed matrix of $Y \in M_\mu(R)$.

We shall also denote by $M_{\mu,\nu}(R)$ the abelian group of matrices with μ rows, ν columns, whose entries belong to R . For $Y \in M_{\mu,\nu}(R)$, we shall denote by ${}_{ij}Y \in R$ the (i,j)-entry of Y . Let us assume that R is a field. For $Y \in M_{\mu,\nu}(R((x)))$, we shall denote by $Y_n \in M_{\mu,\nu}(R)$ the coefficient of x^n in Y , and by ${}_{ij}Y_n \in R$ the coefficient of x^n in ${}_{ij}Y \in R((x))$. For $Y,Z \in M_{\mu,\nu}(R((x)))$, the Hadamard product $Y_*Z \in M_{\mu,\nu}(R((x)))$ is defined by ${}_{ij}(Y_*Z)_n = {}_{ij}Y_n \cdot {}_{ij}Z_n$. Then $(M_{\mu,\nu}(R((x))),+,*)$ is a (non entire) ring with unit; the entries of its unit are $\frac{1}{1-x} \in R((x))$.

DIFFERENTIAL OPERATORS. Differential polynomials in $\partial = x\, d/dx$ (resp. in d/dx) and their coefficients,are denoted by Roman (resp. Greek) letters; e.g. $\Lambda = \frac{1}{\mu!}\frac{d^\mu}{dx^\mu} - \sum_{j=0}^{\mu-1} \gamma_j \frac{1}{j!}\frac{d^j}{dx^j}$.

REFERENCES. Quotations like "cf III(8)", or "theorem IV 5.3". indicate a reference to formula (8) in Chapter III, resp. to the theorem proved in sub-section 5.3 of Chapter IV. When there are several propositions etc... in a single sub-section, they are numbered.

Introduction

This booklet is by itself an introduction, because it is the first one devoted to G-function theory. However this does not mean that G-functions constitute a new topic: they were brought in by C.L. Siegel in 1929, in his famous paper on applications of diophantine approximation. He defined G-functions to be the formal power series $y = \Sigma a_n x^n$ whose coefficients a_n lie in some algebraic number field K, which fulfill the following three conditions:

i) the maximum of the moduli of the conjugates of a_n grows at most Geometrically with n (i.e. is bounded by C^n),

ii) there exists a sequence of natural numbers (d_n) which grows at most geometrically such that $d_n a_m$ is integral for every $m \leqq n$. (i.e. the "common denominator" of $a_0 \dots a_n$ grows at most geometrically with n) ,

iii) y satisfies some linear homogeneous differential equation

$$d^{\mu}/_{dx^{\mu}}\, y + \gamma_{\mu-1}\, d^{\mu-1}/_{dx^{\mu-1}}\, y + \dots + \gamma_0 y = 0$$

with rational function coefficients $\gamma_h \in K(x)$.

After giving some examples, (hypergeometric series $_2F_1$, Abelian integrals ...), Siegel stated some results that he could obtain using the techniques he worked out for so-called E-functions in the same paper, but did not give any detail concerning the proof.

Except for scattered results about particular cases, it was not before forty years later that G-function theory started to develop slowly as a modest chapter of diophantine approximation, in the direction indicated by Siegel. In 1981 a fundamental paper of E. Bombieri appeared, in which not only he proved some of Siegel's irrationality statements in general form (relying on some previous work of A.I. Galoch'kin), but also, and more significantly, he pointed out the local-to-global nature of the theory.

Since then, the theory overflowed out of its original setting, and new connections with arithmetic algebraic geometry appeared (through the works of D.V. and G.V. Chudnovsky, F. Beukers, and the author); a few of them constitute the matter of the present book.

G-functions and differential equations

Meanwhile, point (iii) tended to disappear in the definition of G-functions - maybe because many authors studied components of solutions of linear systems, for which Siegel's definition seems (unduly) insufficient? However this is unfortunate: for instance the (uncountably many) series which satisfy i) and ii) may be quite "pathological", while the (countable) set of G-functions enjoys nice properties, such as the following one (see Chapter VI):

Theorem A Any G-function $y \in K[[x]]$ satisfies $\prod R_v(y) > 0$, where v runs over the places of K such that the radius of convergence $R_v(y)$ of y (considered as a v-adic function) is finite.

Roughly speaking, this means that the v-adic radii of convergence cannot be too small; whether the converse statement holds, under iii), is an interesting open problem (see Chapter V for a partial answer).
In fact, lefting iii) aside in the definition of G-function is more unfortunate, because it sacrifices the geometric nature of Siegel's concept, in light of the following conjecture:

Conjecture G-functions are exactly the solutions in $\bar{\mathbb{Q}}[[x]]$ of Geometric differential equations (over $\bar{\mathbb{Q}}$).

Such a statement is currently believed in by the experts, and our only originality at this point consists in providing a minimal definition of "geometric" differential equations (or polynomials):

Namely they are elements of the multiplicative submonoid of the Weyl algebra $\bar{\mathbb{Q}}[x,d/dx]$ generated by all factors of Picard-Fuchs differential polynomials which control the cohomology of smooth varieties over $\bar{\mathbb{Q}}(x)$ (one can even consider only proper smooth varieties without changing the submonoid, see Chapter II).

One of the main aims in the second part of this book is to prove half of this conjecture, namely:

Theorem B Any solution in $\bar{\mathbb{Q}}[[x]]$ of a geometric differential equation is a G-function.

(See V app.. The difficult case is when 0 is a singularity).

The converse statement seems for the moment to lie beyond the scope of current methods, though some approach already exists via "diagonals" (see Chapter I).

We content ourselves with proving that differential equations satisfied by G-functions share with geometric differential equations very nice p-adic features.

More precisely, let Λ be a differential equation as in point iii) above, and let v be a finite place of K ; then we denote by $R_v(\Lambda)$ the supremum of the real numbers $r \leq 1$ such that Λ admits a full set of solutions, analytic in the v-adic disk of radius r, centered at a "generic" point (see Chapter IV) : for instance, every geometric differential equation has $R_v = 1$ for almost every v (see Chapter V, appendix.) We prove in Chapters IV,V,VI the following result:

Theorem C Let Λ be a differential equation of minimal order, satisfied by a series $y \in K[[x]]$. The following assertions are equivalent:

1) y is a G-function,

2) $\prod_v R_v(\Lambda) > 0$.

The second condition defines what Bombieri calls "Fuchsian differential operator of arithmetic type"; however, for reasons explained in Chapter IV, we shall prefer "G-operator".

Hence we have reduced the above conjecture to a classical conjecture of Bombieri-Dwork, which asserts that G-operators should be "geometric".

The proof of these theorems combines local methods (weak Frobenius structure ...) and global methods (Hermite-Padé approximants, à la Chudnovsky). In fact, we give quantitative results which relate Bombieri's size of y to $\prod R_v(y)$ and $\prod R_v(\Lambda)$. In the same direction, we also compare the algebraic structure of the two sets of functions that the conjecture would identify:

Theorem D G-functions (resp. solutions of geometric differential equations) form a subspace of $\bar{\mathbb{Q}}[[x]]$ which is stable under both usual (= Cauchy) and coefficientwise (= Hadamard) product.

This includes the following fact: if $\Sigma y_n x^n$ satisfies some geometric differential equation, so does $\Sigma y_n^N x^n$ for any positive integer N, whose proof relies heavily upon Hodge theory (degeneration of Leray spectral sequence and semi-simplicity of the monodromy for proper smooth morphisms, see Chapter II).

On the other side, the units in the algebra of all G-functions (under the usual product) are exactly the invertible algebraic functions in $\bar{\mathbb{Q}}[[x]]$. Generalizing a conjecture of Christol, we expect in addition that G-functions whose inverse satisfies condition ii) above (about denominators) are exactly the diagonals of rational (or algebraic, which amounts to the same) functions.

Special values of G-functions

Via theorem B, G-functions become a new tool in arithmetic algebraic geometry thanks to the diophantine theory of their "special values", see chapter VII. The basic result tells that, given G-functions $y_1,\dots,y_\mu$ and a positive integer δ, there exists a constant c ($\leq$ power of $\delta+1$) with the following property: for any non-zero integers a,b such that $|b| \geq c|a|^c$, then any polynomial relation $p(y_1(a/b),\dots,y_\mu(a/b)) = 0$ of degree δ, with coefficients in the base field K, enters as a factor in the specialization at $x = a/b$ of some functional relation $q(y_1,\dots,y_\mu) \equiv 0$ between the y_i's (with coefficients in $K(x)$). In fact, this statement has a many-coloured meaning: indeed, one may understand the symbol $y_i(a/b)$ as the value in the completion K_v taken by the v-adic Taylor series y_i at the point $a/b \in \mathbb{Q} \subset K_v$, for any place v of K such that y_i converges at that point; the constant c does not depend on v. Bombieri has discovered the possibility of handling several, or even all of these places simultaneously, which leads to a sort of "Hasse principle" for values of G-functions. Using theorems

A and C in order to simplify his hypotheses, one may express this Hasse principle as follows, via the notion of a global relation. According to Bombieri, we say that a relation $p(y_1(\xi),\ldots,y_\mu(\xi)) = 0$ is a global (resp. trivial) relation if it holds v-adically for every place v of K for which $|\xi|_v < \mathrm{Min}(R_v(y_1),\ldots,R_v(y_\mu),1)$ (resp. if it comes from a functional relation by specialization at ξ). Then the following finiteness assertion holds true:

Theorem E. Let Ш_δ denote the set of points $\xi \in \bar{\mathbb{Q}}$ where there exists some global non-trivial relation of degree δ at ξ between given G-functions $y_1,\ldots,y_\mu$. Then Ш_δ has bounded height (at most a power of $\delta+1$).

In particular, any subset of Ш_δ of bounded degree over $\mathbb{Q}$ is finite. In fact theorem E is effective; the bound for the height depends only on δ , the size of the y_i's , the order of the differential equations they satisfy, the height and the cardinality of the singular locus of these differential equations. Theorem E or simple experiments show that relations between values of algebraic functions at rational points are "almost never" global. Nevertheless global relations may sometimes be found for some carefully chosen ξ in this special case, and this leads eventually to results of a new kind concerning the diophantine geometry of curves. Let us present here two such results:

Theorem F. Let $y^m = q(z)$ define an irreducible curve C , with $q \in \mathbb{Z}[z]$, monic of degree n . Assume moreover that m and n have a prime common factor $\ell \geq 3$. Then

i) there are only finitely many rational points (y,z) on C such that no prime $\equiv 1 \bmod. \ell$ divides the denominator of z ; in fact, one has the bound $H(z) < 10^{10n^2} H(q)^{8n}$ for any such point (for any polynomial $p \in \mathbb{Q}[z]$, we denote by $H(p)$ the maximum among the absolute values of the numerators and denominators of the coefficients);

ii) there are only finitely many totally real points in $C(\bar{\mathbb{Q}})$ with bounded denominator and degree.

The method of proof of theorem E is a transcendence argument, namely the so-called Gel'fond's method. The same transcendence method, when applied in a different way to series which satisfy properties

i) and ii) in the definition of G-functions, furnishes new criteria of rationality. Before giving an example, let us note that $y = \Sigma a_n x^n$ satisfies i) and ii) iff its size $\sigma(y) := \overline{\lim_{n\to\infty}} \frac{1}{n} h(a_0,\dots,a_n)$ is finite, where h denotes the logarithmic invariant height on the space K^n.

Theorem G. A series $y \in K[[x]]$ is rational iff for every embedding $K \hookrightarrow \mathbb{C}$, y defines a meromorphic function on a complex disk of radius $> \exp(12\sigma(y))$.

In fact, it is possible to give much stronger variants, assuming for instance only a uniformization property (cf chapter VIII), and this leads to Chudnovsky's criterium of algebraicity, from which they deduce a simple effective proof of the isogeny theorem for elliptic curves over $\mathbb{Q}$.

G-functions and periods of algebraic varieties

Let X be a proper smooth variety over $\bar{\mathbb{Q}} \subset \mathbb{C}$. We call "period of X " in degree n , any coefficient divided by $(2i\pi)^n$ of the representative matrix of the canonical isomorphism

$$P_X^n : H_{DR}^n(X) \otimes_{\bar{\mathbb{Q}}} \mathbb{C} \xrightarrow{\sim} H^n(X_{\mathbb{C}}^{an},\mathbb{Q}) \otimes_{\mathbb{Q}} \mathbb{C} \quad ,$$

with respect to bases selected in the algebraic De Rham cohomology $H_{DR}^n(X) := \mathbb{H}^n(X,\Omega_X^{\cdot})$, resp. in the rational singular cohomology of the associated analytic manifold $X_{\mathbb{C}}^{an}$. An element $t \in H_{DR}^{2m}(X)$ is called a Hodge cycle if it lies at the level F^m of the Hodge filtration, and if $(2i\pi)^{-m}P_X^{2m}(t)$ lies in the rational space $H^{2m}(X_{\mathbb{C}}^{an},\mathbb{Q})$. The double rationality feature of Hodge cycles (relatively to the $\bar{\mathbb{Q}}$-space of De Rham cohomology, resp. to the $\mathbb{Q}(2i\pi)$-space of singular cohomology) has the following consequence: every Hodge cycle in $\overset{2m}{\otimes} H_{DR}^n(X) \subset H_{DR}^{2mn}(X^{2n})$ (Künneth) gives rise to polynomial relations with coefficients in $\bar{\mathbb{Q}}(2i\pi)$ among the periods of X (in degree n).

Grothendieck's conjecture. Every polynomial with coefficients in $\bar{\mathbb{Q}}(2i\pi)$ among the periods "comes from Hodge cycles".
(see IX.2 for a more precise statement).
This is known for elliptic curves with complex multiplication

(G.V. Chudnovsky), and for linear relations among periods of any Abelian variety (G. Wüstholz). Nevertheless, Grothendieck's conjecture still remains an outstanding open problem in the case of Abelian varieties. We present here a new approach via G-functions. Indeed, when X varies in a one-parameter family, the periods are given by the values of analytic functions on the base: the "relative periods", which satisfy suitable Picard-Fuchs differential equations. Moreover, expanding the locally invariant relative periods around a "strong degeneration" (see IX 3,4) in Taylor series, one obtains G-functions - in fact diagonals of rational functions - , and it becomes possible to apply the results of the previous paragraph. Making use of results from the theory of variation of Hodge structure, one can prove (IX 5):

<u>Theorem H</u>. Let $X \longrightarrow S$ be an Abelian scheme of relative dimension g over an affine curve S defined over a finite extension K of $\mathbb{Q}$ in $\mathbb{C}$, and let us assume that the fiber of the connected Néron model at some point $s_0 \in (\bar{S}\setminus S)(K)$ is a torus. Let $\delta \geq 0$, and let $s \in S(K)$ be sufficiently close to s_0 in $\bar{S}(\mathbb{C})$ (this proximity condition depends on δ, K, "the" height of s, ... see IX 5). Then every polynomial relation with coefficients in K, of degree $\leq \delta$, between the values at s of the $2g^2$ locally invariant relative periods around s_0, comes from Hodge cycles.
In chapter IX we shall also develop similar results for some projective morphisms more general than Abelian schemes; they apparently fall beyond the range of any other current method.

<u>Global relations among periods</u>

We have seen with theorem E how the existence of global relations leads to much stronger results. Such a favourable situation is encountered in the presence of "exceptional" Hodge cycles in a fiber X_s, for instance when there exist elements of $\mathrm{End}\, X_s$ which do not come from $\mathrm{End}_S X$. In chapter X we shall study a typical case, and prove:

<u>Theorem I</u>. Let X/S be an Abelian scheme as in theorem H. Let us assume in addition that the geometric generic fiber is simple of odd dimension g. Then there are only finitely many fibers X_s ($s \in (\bar{\mathbb{Q}})$) with bounded residual degree $[K(s):K]$, for which there is no ring embedding $\mathrm{End}\, X_s \hookrightarrow M_g(\mathbb{Q})$.

(Note that there does exist an embedding $End_S X \hookrightarrow M_g(\mathbb{Q})$ because of the degeneration at s_0).
For $g > 1$ we believe that this type of result is new (and it is "effective"). For $g = 1$, exceptional fibers X_s are elliptic curves with complex multiplication, $[K(s):K]$ is essentially the class number of the order of complex multiplication, so that the statement is classical: there are only finitely many discriminants with given class number. However our G-function method does not cover this special case - unfortunately, because it would otherwise yield an effective version of Siegel's theorem which links quantitatively discriminant and class number of definite binary quadratic forms!

Vista: global relations and the "mysterious functor"

The failure of the previous method for $g = 1$ and more generally the need for dealing with all periods of X_s and not only the values at s of the locally invariant relative periods, lead one to expand the relative periods no longer at the degeneration s_0 , but instead at some point s_1 of S . This raises at once two problems:

i) the expansions at s_1 are no longer G-functions, but only linear combinations of G-functions, say $y_1,\dots,y_{4g^2}$, with coefficients in the field $K(P_{X_{s_1}})$ generated by the periods of X_{s_1} . The difficulty which arise in constructing special relations among the "Archimedean values" $y_1(s),\dots,y_{4g^2}(s)$ (using periods relations on X_s) is often easily overcome by choosing $K(P_{X_{s_1}})$ as small as possible, e.g. X_{s_1} of CM type.

ii) (most serious) How to construct relations between the p-adic values at s of $y_1,\dots,y_{4g^2}$ when s is p-adically close to s_1 and "exceptional" - for instance when $End\, X_s$ is bigger than $End_S X$?

A natural way of dealing with this problem is by imitation of the Archimedean case. Here the isomorphism P_{X_s} should be replaced by the functorial $P^{(p)}_{X_s}$ obtained by composing Grothendieck's "mysterious isomorphism" which relates the De Rham cohomology to the p-adic etale cohomology, and Artin's isomorphism which links etale and singular cohomology (once a double embedding of the ground number field $K \begin{smallmatrix} \hookrightarrow \mathbb{C} \\ \hookrightarrow \mathbb{C}_p \end{smallmatrix}$ is given). J.M. Fontaine and W. Messing (and later G. Faltings in a more general setting) have indeed constructed this "mysterious" isomorphism (which involves the definition of a p-adic analogue of $2i\pi$), and the associated <u>p-adic periods</u> (which live in $\mathbb{C}_p((2i\pi))$). By functoriality of $P^{(p)}_{X_s}$, non-trivial endomorphisms on X_s lead to period relations, exactly as in the complex case. Unfortunately, the behaviour of $P^{(p)}_{X_z}$ when X_z varies in a family remains rather mysterious. In fact a solution to the above point ii) seems to depend upon the following:

<u>Problem</u>. How can one relate the p-adic periods of X_s to the values at s of the G-functions $y_1,\ldots,y_{4g^2}$? More generally, what are the properties of the mysterious functor with respect to horizontality?

For the applications, the supersingular case is crucial; on the other side, one can raise this problem not only for Abelian schemes.

Anyway, a nice answer would be of importance: aside from giving an effective version of Siegel's theorem as mentioned above, it would also suggest that relations between values at $\xi \in K \subset \mathbb{C}$, say, of solutions $y_1,\ldots,y_n$ in $K[[x]]$ of an absolutely irreducible G-operator Λ (for which 0 is ordinary), have a "tendency" to be factors of global relations, thus providing a large range of applications to theorem E. The heuristic reasons for this are as follows: granting the Bombieri-Dwork conjecture, we may first replace Λ by a Picard-Fuchs equation associated with a proper smooth $K[x]_{(x)}$ - scheme X . The Grothendieck conjecture for the product $X_0 \times X_\xi$ would now show that a relation $q(y_1(\xi),\ldots,y_n(\xi)) = 0$ with coefficients in K comes from Hodge cycles. Deligne's hope states that Hodge cycles should be absolute (see appendix to IX); this would enable us to write similar relations $q_v(y_1(\xi),\ldots,y_b(\xi)) = 0$ which hold v-adically for every Archimedean place v of K for which the v-adic values $y_i(\xi)$ are defined. Furthermore, there is a conjecture of Fontaine which asserts that the "variety of p-adic periods" of X_ξ should be isomorphic to the variety of complex periods (with respect to a double embedding $K \begin{smallmatrix} \hookrightarrow \mathbb{C} \\ \hookrightarrow \mathbb{C}_p \end{smallmatrix}$); in fact, this is a consequence of a consequence of a more general conjecture about the behaviour of absolute Hodge cycles under $P^{(p)}_{X_\xi}$. Together with a favourable answer to the absolute problem, this tends to show that there are corresponding relations q_v at the finite places. On multiplying these (finitely many) relations q_v for all v such that $|\xi|_v < R_v(y_1,\ldots,y_n)$, we would at last obtain a global relation, "containing" the initial relation as a factor.

For other potential applications, we refer to the last section of chapter X. In fact, we believe that the above problem is a key for understanding and resolving a whole hierarchy of arithmetico-geometric problems.

The reader may now skip to the last appendix of the book, where a short and typical application of theorem E is given: a new proof of the transcendence of π .

Here, however, we can only hope that we have given some feeling for the intricate links which relate G-functions to arithmetic Algebraic Geometry.

Part One

What are G-Functions?

Chapter I G-Functions

G-functions appeared in Siegel's paper [56] about diophantine approximation, and led in this context to an extensive literature (see [7] for a small list). In this chapter we present a definition of G-functions (inspired by Bombieri "local-to-global" setting [7]), and define two basic related invariants, namely the size σ (which coincides with Bombieri's one, ibid.) and the global radius. We then turn to examples: rational functions, diagonals, polylogarithms and generalized hypergeometric functions, which we study with some detail; our presentation of diagonals is inspired by Christol [14]. At last we gather some "pathologies".

In the next chapter, we shall explore what should be G-functions (conjecturally).

§ 1. HEIGHTS AND SIZES

1.1 Height of algebraic numbers [45]

Let $\zeta \in \bar{\mathbb{Q}}$ an algebraic number, lying in some number field K. If $\zeta \neq 0$, the following "product formula" holds:

$$\sum_{v \in \Sigma(K)} \log|\zeta|_v = 0 .$$

The (logarithmic absolute) height of ζ is defined to be

$$\sum_{v \in \Sigma(K)} \log^+|\zeta|_v =: h(\zeta) .$$

One has $h(\zeta^r) = |r|h(\zeta)$ for any $\zeta \in K$ and $r \in \mathbb{Q}$. Thanks to our normalizations, $h(\zeta)$ depends only on ζ but not on K. Thus the height is well-defined over $\bar{\mathbb{Q}}$. Let $p = a_0\prod(x-\zeta_i) \in \mathbb{Z}[x]$ the minimal polynomial of ζ over $\mathbb{Z}$. Then the so-called Mahler measure of ζ, defined as $M(\zeta) := |a_0| \prod \mathrm{Max}(1, |\zeta_i|)$, is related to the height via the formula

$$[\mathbb{Q}(\zeta) : \mathbb{Q}]\, h(\zeta) = \log M(\zeta)$$

$$= \int_0^1 \log|p(e^{2\pi t\sqrt{-1}})|dt \text{ (Jensen's formula)}$$

$$= \overline{\lim_{n\to\infty}} \frac{1}{n} \log^+|\mathrm{Resultant}(p, \sum_{i=0}^{n} x^i)|$$

(Langevin's formula).

For a finite family $(A_k)_k$ of matrices, such that all entries belong to K, we set

$$h((A_k)_k) := \sum_{v\in\Sigma(K)} \log^+ \mathrm{Max}_{i,j,k} |{}_{ij}A_k|_v .$$

Once again, this quantity does not depend on the choice of the number field which contains the entries ${}_{ij}A_k$ of the A_k's.

The following classical inequality holds:

$$h(AB) \leq h(A) + h(B) + \log \nu \text{, for any}$$
$$A \in M_{\mu,\nu}(\bar{\mathbb{Q}}),\ B \in M_{\nu,\rho}(\bar{\mathbb{Q}}) .$$

On the other hand, given $A \in M_{\mu,\nu}(K)$ of rank $\mu < \nu$, one can find a non-zero matrix $B \in M_{\nu,1}(\mathcal{O}_K)$ such that $AB = 0$, and

$$h(B) \leq \frac{\mu}{\nu-\mu} (h(A) + \log \nu + c(K)) ,$$

where $c(K)$ depends only on K.

Indeed, taking components relative to some $\mathbb{Q}$-basis of K inside $\mathcal{O}_K$, and using the last displayed formula, one sees that it suffices to handle the case $K = \mathbb{Q}$, where it follows easily from the box principle ("Siegel's lemma", which appeared in the same paper [56]); the point is that A carries

$(\mathbb{Z}_{\substack{\leq n\\ \geq -n}})^{\nu}$ into $(\mathbb{Z}_{\substack{\leq n\nu\|A\|\\ \geq -n\nu\|A\|}})^{\mu}$, so that if

$(2n+1)^{\nu} > (2\nu\|A\|n+1)^{\mu}$, then two distinct elements of $(\mathbb{Z}_{\substack{\leq n\\ \geq -n}})^{\nu}$ have the same image under A , and the difference gives an element of $(\mathbb{Z}_{\substack{\leq 2n\\ \geq -2n}})^{\nu}$ which is killed by A .

1.2 Height of polynomials

Let $Y \in M_{\mu,\nu}(\bar{\mathbb{Q}}[x])$, $Y = \Sigma Y_n x^n$. We write as usual $\deg Y = \operatorname{Max}\{n \,/\, Y_n \neq 0\}$ for $Y \neq 0$. We shall set:

$$h(Y) := (1 + \deg Y)^{-1}\, h((Y_n)_n) .$$

For $\mu = \nu = 1$, it is easy to check that $(1+\deg y)\,h(y) \leq \Sigma(h(\zeta) + \log 2)$, where ζ runs over the roots of y .

1.3 Height of formal power series; G-functions

Let $Y \in M_{\mu,\nu}(\bar{\mathbb{Q}}[[x]])$, $Y = \sum_{n\geq c} Y_n x^n$. We denote by $Y_{\leq N}$ the truncated series $\sum_{n=0}^{N} Y_n x^n \in M_{\mu,\nu}(\bar{\mathbb{Q}}[x])$. We set:

$$h(Y) := \overline{\lim_{N\to\infty}}\, h(Y_{\leq N}) .$$

This is a well-defined quantity in $[0,\infty]$. One checks immediately that this definition reduces to the previous one when Y has only finitely many (actually $\leq 1 + \deg Y$) non-zero coefficients.

DEFINITION. A G-function is a formal power series $y \in \bar{\mathbb{Q}}[[x]]$ whose height $h(y)$ is finite, and which is annihilated by some non-zero element of $\bar{\mathbb{Q}}[x,d/dx]$.

EXPLANATION. This is equivalent to the classical definition (Siegel [56]): $y = \sum_{n\geq 0} y_n x^n \in \bar{\mathbb{Q}}[[x]]$ is a G-function if and only if all the coefficients belong to some fixed number field K , and

i) for every $v \in \Sigma_\infty$; $\sum_{n\geq 0} i_v(y_n)x^n \in \mathbb{C}_v[[x]]$ defines an analytic function around 0 ,

ii) there exists a sequence of natural integers $(d_n)_{n\in\mathbb{N}}$ which grows at most geometrically, such that $d_n y_m \in \mathcal{O}_K$ for $m = o,\ldots,n$,

iii) y satisfies a linear homogeneous differential equation with coefficients in $K(x)$. This equivalence will be proved in 2.3.

1.4 Size of Laurent series

Let $Y \in M_{\mu,\nu}(\bar{\mathbb{Q}}((x)))$, $Y = \sum_{n\geq -N} Y_n x^n$. We set:

$$\sigma(Y) := \begin{cases} 0 & \text{if } Y \text{ is a Laurent polynomial (i.e. if almost all coefficients are } 0\text{)} \\ h(x^N Y) & \text{otherwise .} \end{cases}$$

One checks immediately that this definition depends only on Y , and not on N . The generalization to the case of a finite family of matrices is immediate.

We shall also use constantly the convenient notation:

$$h_{v,n}(Y) := \frac{1}{n} \underset{\substack{i\leq\mu \\ j\leq\nu \\ k\leq n}}{\mathrm{Max}} \log^+ |{}_{ij}Y_k|_v \text{ ; here } v \text{ denotes a place}$$

of some number field K which contains the coefficients ${}_{ij}Y_k$ of the (i,j)-entries of Y for $i \leq \mu$, $j \leq \nu$, $k \leq n$.

However the non-negative real number $\sum_v h_{v,n}(Y)$ does not depend on the choice of K (by the remark made in the index of notations).

LEMMA 1. $\sigma(Y) = \overline{\lim_{n\to\infty}} \sum_v h_{v,n}(Y)$.

Proof: if $Y \in M_{\mu,\nu}(\bar{\mathbb{Q}}[x,1/x])$, we clearly have $\lim_{n\to\infty} \sum_v h_{v,n}(Y) = 0$, so that it is enough to assume that the sequence $(1/_{\varphi(l)})_{l\geq 0}$ of non-zero coefficients of Y is infinite. We then have

$$\sigma(Y) = \overline{\lim_{l\to\infty}}\, 1/_{\varphi(l)} \cdot h(Y_0,\ldots,Y_{\varphi(l)}) = \overline{\lim_{l\to\infty}} \frac{1}{\varphi(l)} \sum_v \operatorname*{Max}_{\substack{i\leq\mu \\ j\leq\nu \\ k\leq\varphi(l)}} \log^+ |{}_{ij}Y_k|_v$$

$$= \overline{\lim_{n\to\infty}} \sum_v \frac{1}{n} \operatorname*{Max}_{\substack{i\leq\mu \\ j\leq\nu \\ m\leq n}} \log^+ |{}_{ij}Y_n|_v \ .$$

□

REMARK. We could everywhere replace the indexing set of summation $\Sigma(K)$ by Σ_f (resp. Σ_∞). Denoting by h_f, σ_f (resp. h_∞, σ_∞) the corresponding notions - finite (resp. infinite) part of the height or size - the above proof shows that $\sigma_f(Y) = \overline{\lim_{n\to\infty}} \sum_{v\in\Sigma_f} h_{v,n}(Y)$. Assume that all coefficients of the entries of Y lie in a fixed number field K . Let d_n the common denominator in $\mathbb{N} \setminus \{0\}$ of the entries of $Y_0,\ldots,Y_n$. One has $\sigma_f(Y) \leq \log \overline{\lim_{n\to\infty}} d_n^{1/n} \leq d\sigma_f(Y)$. The elementary proof is omitted.

LEMMA 2. *Let* $Y \in M_{\mu,\nu}(\bar{\mathbb{Q}}((x)))$.

a) $\operatorname*{Max}_{i,j} \sigma({}_{ij}Y) \leq \sigma(Y) = \sigma(\zeta Y) \leq \sum_{i,j} \sigma({}_{ij}Y)$, *for any* $\zeta \in \bar{\mathbb{Q}}$,

b) $\sigma(d/dx\, Y) \leq \sigma(Y)$, *for any* $n \in \mathbb{N}$,

c) *if the residue* Y_{-1} *of* Y *vanishes*, $\sigma(\int_0^x Y) \leq \sigma(Y) + 1$,

d) *for* $\zeta \in \bar{\mathbb{Q}}$, *set* $Y_{(\zeta)} := \Sigma\, Y_n \zeta^n x^n$. *Then*

$\sigma(Y_{(\zeta)}) \leq \sigma(Y) + h(\zeta)$.

Let $(Y_{[k]})_{k=1}^N$ *a subset of* $M_{\mu,\nu}(\bar{\mathbb{Q}}((x)))$, *then:*

e) $\sigma(\Sigma Y_{[k]}) \leq \sigma((Y_{[k]})_k) \leq \Sigma\sigma(Y_{[k]})$,

f) $\sigma(*Y_{[k]}) \leq \Sigma\sigma(Y_{[k]})$,

g) <u>if</u> $\mu = \nu$, $\sigma(\prod Y_{[k]}) \leq (1+\log N)\sigma((Y_{[k]})_k)$.

<u>Proof</u>: the proof a,b,d,e,f is straightforward, using lemma 1. Let us prove c): by direct computation, we find

$$h_{v,n}(\int_0^x Y) \leq \begin{cases} h_{v,n}(Y) & \text{if } v \in \Sigma_\infty \\ h_{v,n}(Y)+\frac{1}{n}\operatorname{Max}_{m\leq n}\log|m|_v^{-1} & \text{if } v \in \Sigma_f , \end{cases}$$

so that $\sigma(\int_0^x Y) \leq \sigma(Y) + \overline{\lim}\,\frac{1}{n}\log \text{L.C.M.}(1,2,\ldots,n)$, and the inequality c) follows from the prime number theorem. In order to prove g), we use a trick introduced in this context by Shidlovsky (see Galochkin [30], lemma 7). First we assume without loss of generality that $Y_{[k]} \in M_\mu(\bar{\mathbb{Q}}[[x]])$. Let K be the extension of $\mathbb{Q}$ generated by the m first coefficients ${}_{ij}Y_{kl_N}$ of the entries ${}_{ij}Y_{[k]}$ of the $Y_{[k]}$'s , and set $Y = \prod_{k=1}^{N} Y_{[k]}$. We have

${}_{ij}Y_m = \sum_{\Sigma m_k=m}\sum_{l_k=1}^{\mu} {}_{il_1}Y_{1m_1}\,{}_{l_1l_2}Y_{2,m_1}\cdots{}_{l_{N-1}j}Y_{N_1m_N}$. For a finite place $v \in \Sigma_f$, this gives

$$(*)\qquad \log^+|{}_{ij}Y_m|_v \leq \operatorname*{Max}_{\substack{m_1+\ldots+m_N=m\\ i_1,\ldots,i_N,\ j_1,\ldots,j_N}} \sum_{k=1}^{N}\log^+|{}_{i_kj_k}Y_{k_1m_k}|_v .$$

By reordering $Y_1,\ldots,Y_N$, we may suppose that $m_1 \geq m_2 \geq \ldots \geq m_N$, hence $km_k \leq m$. This yields

$$\log^+|{}_{ij}Y_m|_v \leq \sum_{k=1}^{N}\operatorname*{Max}_{m_k\leq m/k}\ \operatorname*{Max}_{i_k,j_k}\log^+|{}_{i_k,j_k}Y_{k,m_k}|_v ,$$ from which

we deduce

$$h_{v,m}(Y) \leq \sum_{k=1}^{N} 1/k\; h_{v,m/k}((Y_{[1]})_1) .$$

For an infinite place $v \in \Sigma_\infty$, we have to add an extra term to the right hand side of (*), namely $\log\#\{m_1,\dots,m_N)/\Sigma m_k = m\} + \log \mu$, which is $o(m)$; in this case we deduce

$$h_{v,m}(Y) \leq \sum_{k=1}^{N} 1/k\; h_{v,m/k}((Y_{[k]})_k + o(1) .$$

By summing over $v \in \Sigma(K)$, we find

$$\sigma(Y) \leq \left(\sum_{k=1}^{N} 1/k\right)\sigma((Y_{[1]})_1) \leq (1 + \log N)\sigma((Y_{[k]})_k) .$$

□

§ 2. RADII

2.1 Local radii of convergence

Let K be a number field, and let $y = \sum_{n\geq 0} y_n x^n \in K[[x]]$. Then for any $v \in \Sigma_K$, $\Sigma\, i_v(y_n)x^n \in \mathbb{C}_v[[x]]$ defines a v-adic Taylor series $y^{(v)}$; we denote by $R_v(y) \in [0,\infty]$ its radius of convergence. By Hadamard's formula, $R_v(y) = \underline{\lim}_{n\to\infty} |y_n|_v^{-1/n}$. More generally, for any Laurent series $y = \sum_{n\geq -N} y_n x^n \in K((x))$, we set $R_v(y) := R_v(x^N y)$; this definition depends only on y but not on N .

2.2 The global radius

For $Y \in M_{\mu,\nu}(K((x)))$, we set

$$\rho(Y) := \sum_v \log^+\left(\operatorname{Min}_{i,j} R_v({}_{ij}Y)\right)^{-1} \in [0,\infty] .$$

LEMMA 1. $\rho(Y) = \sum_v \overline{\lim}_{n\to\infty} h_{v,n}(Y)$; ρ <u>is invariant under finite extension of</u> K .

Proof: Hadamard's formula yields

$$\rho(Y) = \sum_v \operatorname*{Max}_{i,j} \overline{\lim} \frac{1}{n} \log^+|{}_{ij}Y_n|_v = \sum_v \overline{\lim} \operatorname*{Max}_{i,j} \log^+|{}_{ij}Y_n|_v .$$

Thus it is enough to show that

$$\overline{\lim_{n\to\infty}} \frac{1}{n} \operatorname*{Max}_{\substack{i,j\\ m\le n}} \log^+|{}_{ij}Y_m|_v = \overline{\lim_{n\to\infty}} \frac{1}{n} \operatorname*{Max}_{i,j} \log^+|{}_{ij}Y_n|_v .$$

This is a special case, for $t_n = \operatorname*{Max}_{i,j} \log^+|{}_{ij}Y_n|_v$, of the well-known inequality

$$\overline{\lim_{n\to\infty}} \frac{1}{n} \operatorname*{Max}_{m\le n} t_m \le \overline{\lim_{n\to\infty}} \frac{t_n}{n} =: \ell .$$

Indeed, for any $\varepsilon > 0$, let $M_\varepsilon \le N_\varepsilon$ such that $\frac{t_m}{m} \le \ell + \varepsilon$ for $m \ge M_\varepsilon$ and $\frac{t_m}{m} \le \frac{N_\varepsilon}{M_\varepsilon} \ell$ for $m < M_\varepsilon$. Then

$$\frac{1}{n} \operatorname*{Max}_{m\le n} t_m \ge \operatorname{Max}\left(\operatorname*{Max}_{m\le M_\varepsilon} \left(\frac{m}{n}\right) \frac{t_m}{m} , \operatorname*{Max}_{M_\varepsilon \le m \le N_\varepsilon} \left(\frac{m}{n}\right)\frac{t_m}{m}\right) .$$

The second assertion comes readily from the first one.

REMARK. Here again we could replace the indexing set of summation $\Sigma(K)$ by Σ_f (resp. Σ_∞). The above proof yields corresponding formulae $\rho_f(Y) = \sum_{v\in\Sigma_f} \overline{\lim}\, h_{v,n}(Y)$, $\rho_\infty(Y) = \sum_{v\in\Sigma_\infty} \overline{\lim}\, h_{v,n}(Y)$. Furthermore $\rho(Y) = \rho_f(Y) + \rho_\infty(Y)$, and $\sigma_\infty(Y) \le \rho_\infty(Y)$.

LEMMA 2. Let $Y \in M_{\mu,\nu}(K((x)))$.

a) $\operatorname*{Max}_{i,j} \rho({}_{ij}Y) = \rho(Y) = \rho(\zeta Y)$, for any $\zeta \in K$.

b) $\rho(d/dx\ Y) = \rho(Y)$

c) if the residue Y_{-1} vanishes, $\rho(\int_0^x Y) = \rho(Y)$

d) for $\zeta \in K$, $\rho(Y_{(\zeta)}) \le \rho(Y) + h(\zeta)$.

Let $(Y_{[k]})_{k=1}^{N}$ a subset of $M_{\mu,\nu}(K((x)))$.

e) $\rho(\Sigma Y_{[k]}) \leq \rho((Y_{[k]})_k) = \underset{k}{\mathrm{Max}}\ \rho(Y_{[k]})$

f) $\rho({*}Y_{[k]}) \leq \Sigma\rho(Y_{[k]})$

g) if $\mu = \nu$, $\rho(\prod Y_{[k]}) \leq \underset{k}{\mathrm{Max}}\ \rho(Y_{[k]})$.

Proof: straightforward.

□

2.3 We now prove the equivalence stated in 1.3. Let $y \in K[[x]]$. Assume that $h(y) < \infty$. By lemmata 1 of § 1.4 and 2.2, one gets $\rho_\infty(y) < \infty$ and $\sigma_f(y) < \infty$. The first (resp. second) inequality implies condition 1.3 i) (resp. 1.3 ii), taking into account remark 1.4. Conversely, assume that for any $v \in \Sigma_\infty$, $R_v(y) > 0$ (condition 1.3 i) and that $\overline{\lim\limits_{n\to\infty}}\ d_n^{1/n} < \infty$ (condition 1.3 ii) , where d_n denotes the common denominator in $\mathbb{N} \setminus \{0\}$ of $y_0,\dots,y_n$. Then $\sigma(y) \leq \sigma_\infty(y) + \sigma_f(y) \leq \rho_\infty(y) + \log \overline{\lim\limits_{n\to\infty}}\ d_n^{1/n} < \infty$. At last, let $\Lambda = \frac{d^\mu}{dx^\mu} - \sum\limits_{j=0}^{\mu-1} \gamma_j \frac{d^j}{dx^j}$; then $\Lambda(\Sigma y_n x^n) = 0 \Rightarrow \mathbb{Q}(y_0, y_1, \dots)$ is a number field.

□

§ 3. SEVERAL VARIABLES, DIAGONALIZATION

3.1 All what precedes extends in a straightforward manner to the case of elements of $K((\underline{x})) = K((x_1,\dots,x_\nu))$.

For a multi-index $\underline{n} \in \mathbb{N}^\nu$, we denote by $|\underline{n}|$ its length Σn_i ; $\underline{x}^{\underline{n}}$ means $\prod x_i^{n_i}$. Let $y = \sum\limits_{\underline{n}} y_{\underline{n}} \underline{x}^{\underline{n}} \in K((\underline{x}))$; for any place v of K , we set

$$h_{v,n}(y) = \frac{1}{n} \underset{|\underline{k}|\leq n}{\mathrm{Max}} \log^+ |y_{\underline{k}}|_v \ .$$

We also define the global radius (resp. size) by:

$$\rho(y) = \sum_v \overline{\lim_{n\to\infty}}\, h_{v,n}(y)$$

$$\sigma(y) = \overline{\lim_{n\to\infty}} \sum_v h_{v,n}(y) .$$

For $\nu = 1$, previous lemmata show the compatibility with original definitions.

3.2 Diagonalization

One defines the diagonalization map Δ_ν from $K((\underline{x}))$ to $K((x))$ by the formula

$$\Delta_\nu(\Sigma y_{\underline{n}} \underline{x}^{\underline{n}}) = \sum_{n\geq 0} y_{(n,n,\ldots,n)} x^n .$$

This is a useful tool to produce G-functions, through the following lemma (see 4.2):

LEMMA. The following inequalities hold:

$$\rho(\Delta_\nu(y)) \leq \nu\, \rho(y)$$

$$\sigma(\Delta_\nu(y)) \leq \nu\, \sigma(y) .$$

Proof: this follows immediately from the obvious inequality

$$h_{v,n}(\Delta_\nu(y)) \leq h_{v,n\nu}(y) .$$ □

REMARK 1 (Deligne). Assume that for some infinite place v of K, $y^{(v)} := \Sigma\, i_v(y_{\underline{n}})\underline{x}^{\underline{n}}$ is analytic at $\underline{0} \in \mathbb{C}_v^{\nu}$, with $\nu > 1$. Then $\Delta_\nu Y$ is represented by the integral formula

$$(2\pi\sqrt{-1})^{-(\nu-1)} \int_{\substack{|x_2|=\ldots=|x_\nu|=\varepsilon \\ x_1x_2\ldots x_\nu = x}} Y \frac{dx_2\ldots dx_\nu}{x_2\ldots x_\nu}$$

for ε and $|x|$ small enough.

This follows from the residue formula:

$$(2\pi\sqrt{-1})^{-(\nu-1)} \int_{\substack{|x_2|=\ldots=|x_\nu|=\varepsilon \\ x_1x_2\ldots x_\nu = x}} \underline{x}^{\underline{n}} \frac{dx_2\ldots dx_\nu}{x_2\ldots x_\nu} \begin{cases} = x^{n_1} & \text{if } n_1=n_2=\ldots n_\nu \\ =0 & \text{otherwise.} \end{cases}$$

REMARK 2. It seems that diagonals were first introduced in the study of Hadamard products (see e.g. [9]). This relationship is given by the formula:

$$\Delta_\nu(y_1(x_1)\ \dots\ y_\nu(x_\nu)) = y_1 * \dots * y_\nu \ .$$

3.3 Geometric interpretation

Let us set $W = \operatorname{Spec} K(x)[\underline{x}]/_{(x_1 x_2 \dots x_\nu - x)}$, with $\nu > 1$. Let (E,∇) be a coherent module with integrable connection over some affine open subset U of W , and let θ be some horizontal $K(U)$-linear map from E to $K((\underline{x}))$; in other words, $y := \theta(e)$, for $e \in \Gamma E$, is a solution in $K((\underline{x}))$ of an "integrable differential equation".

We consider the $K(x)$-linear map:

$$\Delta_{\nu,\theta} : e \otimes \frac{dx_2 \dots dx_\nu}{x_2 \dots x_\mu} \longmapsto \Delta_\nu(\theta(e)) \ , \text{ for all local sections } e \text{ of } E \ .$$

PROPOSITION. <u>The map</u> $\Delta_{\nu,\theta}$ <u>induces a horizontal map from the algebraic De Rham cohomology group</u> $H^{\nu-1}_{DR}(U,(E,\nabla))$ <u>endowed with Gauss-Manin connection relative to</u> $K(x)$ (see [39]), <u>to</u> $K((x))$ <u>endowed with exterior derivative</u>.

<u>Proof</u>: the smooth scheme U is affine, thus there is an isomorphism $H^{\nu-1}_{DR}(U,(E,\nabla)) \simeq E_U \otimes \Omega^{\nu-1}_{U/K(x)} \Big/ \nabla_{\nu-1}(E_U \otimes \Omega^{\nu-2}_{U/K(x)})$,

where the value at d/dx of the Gauss-Manin connection acts through $\nabla(d/_{d(x_1 x_2 \dots x_\mu)})$ on E . The statement would follow from Deligne's integral formula if $\theta(e)^{(v)}$ were analytic at $\underline{0}$ for some $v \in \Sigma_\infty$. However this can fail if $\underline{0}$ corresponds to an irregular singularity of (E,∇) ; thus we shall rather translate a purely algebraic argument from Christol [14]. The relation $\Sigma \frac{dx_i}{x_i} = 0$ in $\Omega^1_{W/K(x)}$, together with the formula

$$\Delta_\nu\left(x_i \frac{\partial \theta(e)}{\partial x_i}\right) = x \frac{d}{dx} \Delta_\nu(\theta(e)) \ , \text{ yields}$$

$$\Delta_{\nu,\theta}\Big(\nabla_{\nu-1}\Big(e\otimes \frac{dx_2\dots\widehat{dx_i}\dots dx_\nu}{x_2\dots\widehat{x_i}\dots dx_\nu}\Big)\Big)=\Delta_{\nu,\theta}\Big(\big(x_i\nabla(\partial/\partial x_i)e-x_1\nabla(\partial/\partial x_1)e\big)\otimes \frac{dx_2\dots dx_i\dots dx_\nu}{x_2\dots x_i\dots x_\nu}\Big)$$

$$=\Delta_\nu\left(x_i\,\frac{\partial\theta(e)}{\partial x_i}-x_1\,\frac{\partial\theta(e)}{\partial x_1}\right)=0\ .$$

Therefore $\Delta_{\nu,\theta}$ factors through $H^{\nu-1}_{DR}(U,(E,\nabla))$. In order to prove the horizontality statement, we fix $x_2,\dots,x_\nu$ and get

$$\Delta_{\nu,\theta}\Big(x_1\nabla(\partial/\partial x_1)e\otimes \frac{dx_2\dots dx_\nu}{x_2\dots x_\nu}\Big)=\Delta_\nu\Big(x_1\,\frac{\partial\sigma(e)}{\partial x_1}\Big)=x\; d/dx\;\Delta_{\nu,\theta}\Big(e\otimes \frac{dx_2\dots dx_\nu}{x_2\dots x_\nu}\Big)\ .$$

□

COROLLARY 1. *Assume that* $H^{\nu-1}_{DR}(U,(E,\nabla))$ *is finite-dimensional over* $K(x)$ (*assume for instance that* (E,∇) *has only regular singular points*, see next chapter, § 2.2) *then for* $y=\theta(e)$ *as above*, $\Delta_\nu(y)$ *satisfies an ordinary linear homogeneous differential equation with coefficients in* $K(x)$.

□

COROLLARY 2. *Assume that* θ *is a solution in* $K((\underline{x}))$ *of the Picard-Fuchs system* $H^{\mu}_{DR}(Y/K(\underline{x}))$ *of a smooth proper* $K(\underline{x})$*-variety* Y . *Then* $\Delta_{\nu,\theta}$ *is a solution in* $K((x))$ *of the Picard-Fuchs system* $H^{\mu+\nu-1}_{DR}(Z/K(x))$ *of a smooth* $K(x)$*-variety* Z .

Proof: let V be an open dense subset of $\operatorname{Spec} K\left[\underline{x},\frac{1}{x_1\dots x_\nu}\right]$ such that Y extends to a smooth proper morphism $Y_V \xrightarrow{f} V$, and let us denote by g the obvious smooth morphism $V \longrightarrow \operatorname{Spec} K\left[x_1\dots x_\nu,\ \frac{1}{x_1\dots x_\nu}\right]$. Let us consider the Cartesian squares:

$$\begin{array}{ccc}
Z & \longrightarrow & Y_V \\
\downarrow & & \downarrow f \\
U & \longrightarrow & V \\
\downarrow & & \downarrow \\
W & \longrightarrow & \operatorname{Spec} K\left[\underline{x},\ \frac{1}{x_1\dots x_\nu}\right]
\end{array}\ .$$

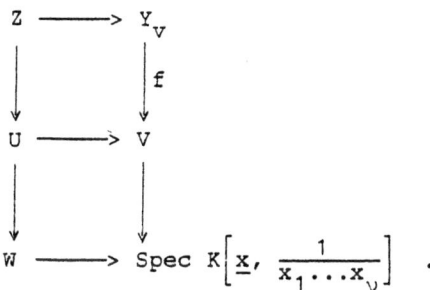

According to the proposition, $\Delta_{\nu,\theta}$ is a solution in $K((x))$ of $H_{DR}^{\nu-1}(U/K(x), H_{DR}^{\mu}(Z/U))$.

On the other hand, there is the Leray spectral sequence

$$(*) \qquad H_{DR}^{\nu-1}(U/K(x), H_{DR}^{\mu}(Z/U)) \Rightarrow H_{DR}^{\mu+\nu-1}(Z/K(x)) .$$

Let us extend the scalars K to $\mathbb{C}$; since $f_{\mathbb{C}}$ is proper and smooth, the Leray spectral sequence of local systems $R^{\nu-1}g_{\mathbb{C}*} R^{\mu}f_{\mathbb{C}*}(\mathbb{C}) \Rightarrow R^{\mu+\nu-1}(gf_{\mathbb{C}})_*(\mathbb{C})$ degenerates [24] 2.4. It follows from the comparison theorem that (*) also degenerates as a spectral sequence of $K(x)$-vector spaces with connection. Thus $\Delta_{\nu,\theta}$ is a solution of $H_{DR}^{\mu+\nu-1}(Z/K(x))$.

□

REMARK 3. Combining corollary 2 with remark 2, we get that if $\Sigma a_n x^n$ satisfies a Picard-Fuchs equation from projective geometry, then for any N , $\Sigma a_n^N x^n$ satisfies a Picard-Fuchs equation.

§ 4. EXAMPLES

We shall study four typical classes of G-functions, each of which is stable under Hadamard product; namely: rational functions, diagonals of rational functions in several variables, polylogarithms and hypergeometric functions (Geometric and hypergeometric series were already put forward by C.L.Siegel [56], and G-functions borrow their generic name from these special cases). Each of these series satisfies some linear homogeneous differential equation, which turns out to come from geometry.

4.1 Rational functions

Let $y \in K(x)$, and let us write $pol(y)$ for the set of poles of y . We may write y as the quotient p/q of two polynomials in $\mathcal{O}_K[x]$. Let us write N for the norm of the first non-zero coefficient of q ; then $y \in \mathcal{O}_K[1/N]((x))$. On the other hand, it is immediate that $\rho_\infty(y) < \infty$. Since such series occur frequently, we state a

DEFINITION (Christol). A Laurent series $y \in K((x))$ is globally bounded if and only if

i) for any $v \in \Sigma(K)$, $R_v(y) > 0$,

ii) there exists $N \in \mathbb{N}^{\times}$ such that $y \in \mathcal{O}_K[1/N]((x))$.

LEMMA. Any $y \in K(x)$ satisfies $\rho(y) = \sigma(y) = h(\mathrm{pol}(y))$.

Proof: we have $R_v(y) = \operatorname*{Min}_{\zeta \in \mathrm{pol}(y)} |\zeta|_v$ for any $v \in \Sigma(K)$, whence the equality $\rho(y) = h(\mathrm{pol}(y))$.

On the other side, the fact that y is globally bounded implies that $h_{v,n}(y) = 0$ for almost all v, and all n. Using lemmata 1 of §§ 1.3 and 2.2, we come by the inequality $\sigma(y) \leqq \rho(y)$. In order to show that it is an equality, it suffices to establish the existence of the limit $\lim_{n\to\infty} h_{v,n}(y)$ for any $v \in \Sigma(K)$; but this follows from the fact that the coefficients of y satisfy linear recurrence equations for $n >> 0$ (see remark 2 below), or by decomposition into simple elements.

□

REMARK 1. This lemma, together with the identity $d/dx(p/q) = (p'/p - q'/q)\cdot p/q$ shows that rational functions are G-functions.

REMARK 2. The lemma generalizes immediately to the case of a matrix $Y \in M_{\mu,\nu}(K(x))$. The stability of $M_{\mu,\nu}(K(x))$ under Hadamard product is easily seen using the characterization of rational series: $y \in K(x) \iff \exists\, N \in \mathbb{N}^{\times}$, $\exists\, Y,Z \in M_N(K)$ such that $Y_n = \operatorname{tr} YZ^n$ (existence of recurrence relations); we have the formula $(Y_1 * Y_2)_n = \operatorname{tr}(Y_1 \otimes Y_2)(Z_1 \otimes Z_2)^n$, with obvious notations.

4.2 Diagonals of rational functions

We shall denote by $K[\underline{x}]_{(\underline{x})}$ the localization of the ring $K[\underline{x}] = K[x_1,\dots,x_\nu]$ at the ideal generated by $x_1,\dots,x_\nu$, and by $K\{x\}$ the henselization of $K[x]$ at the ideal generated by x (i.e. the subring of $K[[x]]$ of algebraic elements over $K(x)$.)

DEFINITION. Elements in the target $\Delta_\nu(K[\underline{x}]_{(\underline{x})})$ of the diagonalization map restricted to $K[\underline{x}]_{(\underline{x})}$ are called diagonals of rational functions (over K).

REMARK 1. Let us consider again the geometric interpretation of Δ_ν in § 3.3. In the present case, let $p/q \in K[\underline{x}]_{(\underline{x})}$, with $p,q \in K[\underline{x}]$. We may take for U the subset of X where q does not vanish; $E = \mathcal{O}_U$, endowed with exterior derivative ∇ ; θ : the standard horizontal map $\mathcal{O}_U \longrightarrow K((\underline{x}))$, where x is replaced by $x_1x_2\dots x_\nu$; $e := p/q$. We have $H^{\nu-1}_{DR}(U,(E,\nabla)) = H^{\nu-1}_{DR}(U)$, the ordinary algebraic De Rham cohomology of the smooth affine scheme U . This is a finite-dimensional $K(x)$-vector space; see [49] for an algebraic proof which does not use resolution of singularities. According to corollary 3.3, diagonals of rational functions satisfy "Picard-Fuchs" differential equations associated to smooth affine $K(x)$-schemes.

LEMMA. Let $y \in K[[x]]$, $y = \Delta_\nu(p/q)$ be a diagonal of rational function. Then y is a globally bounded G-function, and $\sigma(y) \leq \rho(y) < \infty$.

Proof: we may assume that $p,q \in \mathcal{O}_K[\underline{x}]$; let us denote by N the norm of $q(0) \neq 0$. Then it is clear that $p/q \in \mathcal{O}_K$ and $y \in \mathcal{O}_K[1/N][[x]]$. On the other side, the v-adic radius of convergence $R_v(P/q)$ is non zero for every $v \in \Sigma(K)$, and the same holds for $R_v(y)$ according to Hadamard's formula. Using the last remark, this shows that y is a globally bounded G-function. The deduction $\sigma(y) \leq \rho(y)$ is made as in lemma 4.1. In fact, it could be shown that $\sigma_f(y) = \rho_f(y) \leq \nu h_f(q(\underline{0})^{-1}) \leq \nu\, h(q(\underline{0}))$.

It happens that diagonalsof rational functions occur very frequently, even though it is often difficult to find the (non-unique) relevant rational function. To explain this fact,

G. Christol [15] has set the following conjecture up:

CONJECTURE. Every globally bounded solution in $K[[x]]$ of a linear homogeneous differential equation with coefficients in $K[x]$ is the diagonal of some rational function.

In other words, every globally bounded G-function should be a diagonal. We now prove that algebraic functions are diagonals of rational functions in two variables (Christol-Furstenberg [10][29]. Consequently, they are globally bounded (Eisenstein).

PROPOSITION. The equality $\Delta_2(K[x_1,x_2]_{(x_1,x_2)}) = K\{x\}$ holds.

Sketch of proof: in fact we shall only consider the inclusion $\supseteq$. Let $y \in K\{x\}$ and let $r(y,x) := 0$ be a polynomial equation for y. Assuming that $r(0,0) = 0$, $\left.\frac{\partial r}{\partial y}\right|_{(0,0)} \neq 0$, $\left.\frac{\partial r}{\partial x}\right|_{(0,0)} \neq 0$, we shall exhibit a rational function p/q such that $\Delta_2(p/q) = y$. We set $q(x_1,x_2) = \frac{1}{x_1} r(x_1, x_1x_2)$, so that $1/q \in K[x_1,x_2]_{(x_1,x_2)}$, and $\left.\frac{\partial q}{\partial x_2}\right|_{(0,0)} \neq 0$.

Let us consider the following diagram (where W and U have the same meaning as in remark 1, and $Z = W \setminus U$):

$$\begin{array}{ccccccccc}
0 & \xrightarrow{=} & H^1_{DR}(W \cup \{0\}) & \longrightarrow & H^1_{DR}(U) & \xrightarrow{\mathrm{Res}_{Z\cup\{0\}}} & H^0(Z\cup\{0\}) & \longrightarrow & 0 \\
 & & \downarrow & & \| & \nwarrow \overset{\varphi}{\cdots} & \uparrow & & \\
0 & \longrightarrow & H^1_{DR}(W) & \longrightarrow & H^1_{DR}(U) & \xrightarrow{\mathrm{Res}_Z} & H^0(Z) & \longrightarrow & 0
\end{array}$$

where all arrows are horizontal maps, and where the horizontal rows are the residue exact sequences: Res_Z is the "coefficient of dq/q", given at the stage of differential forms by

$$\mathrm{Res}_Z\left(p/q \; {dx_2}/{x_2}\right) = \left(\frac{\partial q}{\partial x_2}\right)^{-1} p/x_2 \Big|_{q(x_1,x_2)=0} .$$

Now the derivation d/dx extends in a unique way to $K(x,y)$, whence a connection on this space, which can be identified with the Gauss-Manin connection on $H^0(Z)$. It follows that the image of $y \in K(x,y) \simeq H^0(Z)$ under φ is given by the class of $p/x \cdot dx_2/x_2$ where $p = x_1 x_2\, \partial q/\partial x_2$.

The following diagram of horizontal maps

$$\begin{array}{ccccc} H^1_{DR}(U) & \xleftarrow{\quad\varphi\quad} & H^0(Z) & \xleftarrow{\quad\approx\quad} & K(x,y) \\ {\scriptstyle \Delta_{2,\theta}}\downarrow & & & & \downarrow \\ K[[x]] & & = & & K[[x]] \end{array}$$

(where θ is defined in the above remark) shows that $(\Delta_{2,\theta}\circ\varphi)(Y)$ satisfies the same differential equation as y, and $(\Delta_{2,\theta}\circ\varphi)(y)\Big|_0 = x\,\Delta_2\left(\frac{1}{q}\,\partial q/\partial x_2\right)\Big|_0 = 0$. It follows that
$$y = \Delta_2\left(\frac{x_1 x_2}{q}\cdot \partial q/\partial x_2\right).$$ □

For a proof of the reversed inclusion $\subseteq$, with an argument from linguistics, see [28] 5.

REMARK 3: the stability of diagonals of rational functions under Hadamard product is immediate from the formula:

$$\Delta_{\nu_1+\nu_2}(r_1(x_1,\ldots,x_{\nu_1})\, r_2(x_{\nu_1+1},\ldots,x_{\nu_1+\nu_2})) = \Delta_{\nu_1} r_1 * \Delta_{\nu_2} r_2 .$$

However the subclass of algebraic functions is not stable under $*$; by way of counterexample, one may take (Jungen, 1931):

$$(1-x)^{1/2} * (1-x)^{-1/2} = \Delta_4(4/(2-x_1-x_2)(2-x_3-x_4)) = {}_2F_1(1/2,1/2,1,x)$$
$$= \sum_{n\geq 0} \binom{2n}{n}^2 \left(x/16\right)^n, \text{ which is transcendental.}$$

4.3 Polylogarithms

We turn back to more down-to-earth examples. Let $L_k = \sum_{n\geq 0} x^n/n^k$ be the kth-polylogarithmic series. It satisfies the "unipotent" differential equation: $d/dx\, \frac{1-x}{x}\,(xd/dx)^k\, L_k = 0$ obtained from the chain rule $xd/_{dx} L_k = L_{k-1}$, $L_0 = x/_{1-x}$; the other solutions can be expressed by means of the functions $1, \log x, \dots, \log^{k-1} x$; the singularities are $0, 1, \infty$.

LEMMA. One has $\rho(L_k) = 0$, $\sigma(L_k) = k$.

Proof: this is a straightforward consequence of the prime number theorem. □

Moreover, one can show that $\overline{\lim_{k\to\infty}} \sigma(L_1^k)/_{\log k} = 1$ (cf. Lemma 2g)); see ch. VIII.

REMARK. Integration of any formal power series y is nothing but the Hadamard product $xy * L_1$.

4.4 Generalized hypergeometric functions

For $a \in \mathbb{Q}$, we set $(a)_0 = 1$, $(a)_{n+1} = (a+n)(a)_n$, and for $\underline{a} := (a_1, \dots, a_\mu) \in \mathbb{Q}^\mu$ we set $(\underline{a})_n = \prod_{m=1}^{\mu} (a_m)_n$. To any couple $(\underline{a}, \underline{b})$ in $(\mathbb{Q}-\{-\mathbb{N}\})^\mu \times (\mathbb{Q}-\{-\mathbb{N}\})^\nu$, we associate the hypergeometric function

$$y = F(\underline{a}, \underline{b}, x) := \sum_{n\geq 0} {(\underline{a})_n}/{(\underline{b})_n} \, x^n .$$

LEMMA. The three conditions $\rho(y) < \infty$, $\sigma(y) < \infty$ and $\mu = \nu$ are equivalent. If they are satisfied, one has

$$\rho(y) = \sigma(y) \leq \operatorname{Max}\left(\sum_{m=1}^{\mu} (2h_f(a_m) - h_f(b_m)), 0 \right) .$$

Proof: either of the conditions $\rho(y) < \infty$, $\sigma(y) < \infty$ implies that for $v \in \Sigma_\infty$, $R_v(y) > 0$ which implies in turn that $\mu \leq \nu$, and $R_v(y) \geq 1$ (hence $\rho_\infty(y) = \sigma_\infty(y) = 0$). Let N be the greatest common denominator of the a_m, b_m's ; for $p > N$ and $n \to \infty$, we have

$$\left| {(a_m)_n}/{(b_m)_n} \right|_p = O\left(p^{\log n}\right) ,$$

$$\left| {1}/{(b_n)_n} \right|_p^{1/n} \sim p^{1/p-1} ,$$

$$\operatorname{den}\left(N^n (a_m)_n/(b_m)_n \right) = O\left(e^{1/\log n}\right) ,$$

and $\left(\text{den } N^n/(b_m)_n\right)^{1/n} \sim n/e$ /Stirling), see the appendix. The former two estimates, together with the divergence of $\sum_{p>N} \frac{\log p}{p-1}$, show that $\rho(y) < \infty \Rightarrow \mu \geq \nu$.

The latter two estimates show that $\sigma(y) < \infty \Rightarrow \mu \geq \nu$. Conversely the first and the third estimates show that $\mu = \nu$ implies finiteness for ρ and σ, and that

$$\rho(y) = \sum_{p|N} \overline{\lim_{n\to\infty}}\, h_{p,n}$$

$$\sigma(y) = \overline{\lim_{n\to\infty}} \sum_{p|N} h_{p,n} .$$

A straightforward computation (remarking that $|(a_m)_n|_p = |a_m|_p^n$ if $|a|_p > 1$) then leads to the inequality

$$\rho(y) = \sigma(y) \leq \operatorname{Max}\left(\sum_{m=1} (2\log \text{den } a_m - \log \text{den } b_m), 0\right) . \qquad \square$$

REMARK 1. We could define hypergeometric series for parameters $(\underline{a},\underline{b})$ in $(K\setminus\{-\mathbb{N}\})^{\mu+\nu}$ for any number field. However it follows from methods of chapter VI that such a hypergeometric series is a G-function only if $(\underline{a},\underline{b}) \in (\mathbb{Q}\setminus\{-\mathbb{N}\})^{\mu+\nu}$, see VI ex.1.

REMARK. G. Christol [15] has determined all globally bounded hypergeometric functions. The extra condition is the following one: let N as above; then for any M with $0 \leq M < N$ and $(M,N) = 1$, and for any positive integer j with $j \leq \mu$, $\#\{i/Ma_i \,\alpha\, Mb_j\} \geq \#\{i/Mb_i \,\alpha\, Mb_j\}$ (here α is the total ordering of $\mathbb{R}$ defined by

$$y \,\alpha\, z \Longleftrightarrow y + [-y] < z + [-z] \quad \text{or} \quad (y + [-y] = z + [-z] \text{ and } y \geq z)).$$

Let us now introduce the classical Meijer G-functions, which however are <u>not</u> G-functions in Siegel's sense! These are integrals of Mellin-Barnes type over a suitable loop:

$$G^{m,n}_{\nu,\mu}(\underline{a},\underline{b},x) := \frac{1}{2\pi\sqrt{-1}} \oint \frac{\prod_{j=1}^{m} \Gamma(b_j - s) \prod_{j=1}^{m} \Gamma(1-a_j+s)}{\prod_{j=m+1}^{\mu} \Gamma(1-b_j+s) \prod_{j=n+1}^{\nu} \Gamma(a_j - s)} x^s ds ,$$

$$\text{for}\quad 0 \le m \le \mu,\ 0 \le n \le \nu .$$

In the case $\mu = \nu$, these functions satisfy some Fuchsian differential equation. Namely, $z := G^{m,n}_{\mu,\mu}(\underline{a},\underline{b},(-1)^{m+n}x)$ satisfies the equation

$$(*)\qquad (-1)^{\mu}x \prod_{j=1}^{\mu}(\partial-a_j+1)z = \prod_{j=1}^{\mu}(\partial-b_j)z \quad \text{where} \quad \partial = x\, d/dx ,$$

whose singularities are $x = 0,\ (-1)^{\mu}$ and ∞.

The link with hypergeometric series is given by the formulae

$$F(\underline{a},\underline{b},x) = \frac{\prod_{j=1}^{\mu}\Gamma(b_j)}{\prod_{j=1}^{\mu}\Gamma(a_j)} G^{\mu,1}_{\mu,\mu}(\underline{a},\underline{b},-1/x) = \frac{\prod_{j=1}^{\mu}\Gamma(b_j)}{\prod_{j=1}^{\mu}\Gamma(a_j)} G^{1,\mu}_{\mu,\mu}(\underline{1}-\underline{a},\underline{1}-\underline{b},x)$$

and

$$G^{m,n}_{\mu,\mu}(\underline{a},\underline{b},x) = \sum_{k=1}^{m} \frac{\prod_{\substack{j=1\\ j\neq k}}^{m}\Gamma(b_j-b_k)\prod_{j=1}^{n}\Gamma(1+b_k-a_j)}{\prod_{j=m+1}^{\mu}\Gamma(1+b_k-b_j)\prod_{j=n+1}^{\mu}\Gamma(a_j+b_k)}\, x^{b_k}\, F(\underline{-a}+\underline{1+b_k},\underline{-b}+\underline{1+b_k},(-1)^{\mu-m-n}x)$$

where we set $\underline{h} = (h,\dots,h)$ for any $h \in \mathbb{Q}$, see [4] 5.5. The latter formula shows that $G^{m,n}_{\mu,\mu}$ is a linear combination (with transcendental constant coefficients) of some Siegel G-functions

REMARK 3. In the case $\mu = \nu = 1$, we have $F(a,b,x) = {}_2F_1(a,1,b,x)$, the classical hypergeometric function, and it is well-known that equation (*) is a factor of a Picard-Fuchs equation [40]. For higher μ, this is by no means obvious. However it remains that:

PROPOSITION. (for $\mu = \nu$) $F(\underline{a},\underline{b},x)$ <u>satisfies some Picard-Fuchs differential equation</u>.

<u>Proof</u>: according to remarks of § 4.2, we have

$$F(\underline{a},\underline{b},x) = \mathop{*}_{i=1}^{\nu}\left({}_2F_1(a_i,1,b_i,x)\right) = \Delta_{\nu}\left(\prod_{i=1}^{\nu}{}_2F_1(a_i,1,b_i,x_i)\right) .$$

By corollary 2 in § 3.3, it suffices to show that $\prod_{i=1}^{\nu} {}_2F_1(a_i,1,b_i,x_i)$ satisfies a Picard-Fuchs differential equation associated $H^{\mu'}_{DR}(Y/_{\mathbb{Q}(x)})$ for some proper smooth Y. Using Künneth formula in algebraic De Rham cohomology, it is enough to prove this statement for $\nu = 1$. If $b \in \mathbb{N}^{\times}$, then ${}_2F_1(a,b,x)$ is algebraic and the statement holds with $\mu^1 = 0$. If $b \notin \mathbb{N}^{\times}$, we use Gauss relations between contiguous hypergeometric series

$$(b-a-1)\,{}_2F_1(a,1,b,x)+a\,{}_2F_1(a+1,1,b,x)-(b-1)\,{}_2F_1(a,1,b-1,x)=0$$

$$b[a-(b-1)x]\,{}_2F_1(a,1,b,x)+ab(1-x)\,{}_2F_1(a+1,1,b,x)+(b-1)(b-a)x\,{}_2F_1(a,1,b+1,x)=0$$

in order to reduce ourselves to the case $a > 0,\ 1 > b > 2$. In this case, Euler's integral representation ${}_2F_1(a,1,b,x) = (b-1)\int_0^1 (1-t)^{b-2}(1-tx)^{-a}dt$ shows that ${}_2F_1(a,1,b,x)$ satisfies the Picard-Fuchs equation associated to the differential $\frac{dt}{u}$ over the smooth completion of the curve

$$u^N = (1-t)^{(2-b)N}(1-tx)^{aN}\ ,\quad N = \mathrm{den}(a,b)\ . \qquad \square$$

§ 5. COUNTEREXAMPLES

In this paragraph, we gather some "pathological" examples to show that there is no link in general between ρ and σ. We shall show later that for solutions of linear homogeneous differential equations with coefficients in $\bar{\mathbb{Q}}(x)$, ρ and σ are in contrast closely related. We also state that ρ and σ are bad-behaved under inversion of functions.

5.1 <u>A G-function whose inverse satisfies</u> $\rho = \sigma = \infty$.
Recall that $\rho(L_1/x) = 0,\ \sigma(L_1/x) = 1$. Let $y = x/L_1$, so that $y_0 = 1$ and $y_n = \sum_{m=1}^{n} \frac{y_{n-m}}{m+1}$. For each pth root of unity $\zeta \in \mathbb{C}_p \setminus \{1\}$, L_1/x vanishes at $1-\zeta$, and $|1-\zeta|_v = |p|_v^{1/p-1}$. Therefore $R_v(y) \leq |p|_v^{1/p-1}$ and

$\rho(y) = \infty$. It can be shown (see VIII ex.1) that $\sigma(y)=\infty$; it will follow that the composite series $L_1 \circ L_1 \in \mathbb{Q}[[x]]$ has infinite size too, since

$$x(1-x)\ d/dx\ (L_1 \circ L_1) = y\ .$$

5.2 <u>An example with</u> $\rho = 0$ <u>and</u> $\sigma = \infty$

We set $y = \sum_{k\geq 1} k^{-[k/\log^2 k]} x^k$. We readily compute

$$h_{p,n}(y) = \begin{cases} 0 \quad \text{for} \quad p = \infty \\ \frac{1}{n} \operatorname*{Max}_{k\leq n} [{}^{k}/\log^2 k][\log k/_{\log p}]\log p = o_n(1)\ , \\ \qquad \text{for} \quad p \quad \text{a finite prime.} \end{cases}$$

Thus $\overline{\lim\limits_{n\to\infty}}\, h_{p,n}(y) = 0$ and $\rho(y) = 0$. On the other side

$$\sum_{p \text{ prime} \neq \infty} h_{p,n}(y) = \frac{1}{n} \log \operatorname*{l\cdot c\cdot m}_{k\leq n} (k^{[k/\log^2 k]})$$

$$\geq \frac{1}{n} \sum_{p\leq n} (p/\log p - \log p) \longrightarrow \infty \quad \text{when} \quad n \to \infty\ .$$

This shows that $\sigma(y) = \infty$.

5.3 <u>An example with</u> $\rho = \infty$ <u>and</u> σ <u>arbitrarily small</u>.

Let $N \geq 0$ and let us set

$$y = \sum_{\substack{p \text{ prime} \\ \neq\infty}} \sum_{k\geq 0} p^{-[2^{p^k - N}/\log p]} \cdot x^{p\cdot 2^{p^k}}\ .$$

We have $h_{p,n}(y) = \begin{cases} 0 \quad \text{for} \quad p = \infty \\ [2^{\{n,p\}-N}/\log p]\ \frac{\log p}{n} \quad \text{for any finite prime } p\ , \end{cases}$

denoting by $\{n,p\}$ the maximal power of p such that $2^{\{n,p\}} \leq n/p$. Thus $\overline{\lim\limits_{n\to\infty}}\, h_{p,n}(y) = 2^{-N}/p$ in the latter case,

and $\rho(y) = \infty$. Now we have

$$\Sigma h_{p,n}(y) = \frac{1}{n} \sum_{p \leq n} [2^{\{n,p\}-N}/\log p] \log p$$

$$\leq \frac{1}{n} \sum_{p \leq n} 2^{\{n,p\}-N} .$$

We note that for $p \neq q$, then $\{n,p\} \neq \{n,q\}$, so that

$\sum_{p \leq n} 2^{\{n,p\}} \leq 2^{\{n,p_0\}} \sum_{k=1}^{\infty} 2^{-k}$ for some p_0, $2 \leq p_0 \leq n$.

Therefore $\sigma(y) \leq 2^{-N}$. Examples with $\rho = \infty$ and $\sigma = 0$ do not exist (ex.1).

5.4 A globally bounded function with $\sigma < \rho$

Let us consider

$$y = \sum_{k \geq 0} 2^{(-2)^k} x^{2^k} .$$

We have
$$h_{p,n}(y) = \begin{cases} 0 & \text{for } p \neq 2,\ p \neq \infty \\ 2^{2\left[\frac{1}{2}\left[\frac{\log n}{\log 2}\right]\right]} \log 2 , & \text{for } p = 2 \\ \left[\frac{-1}{2}\left[\frac{\log n}{\log 2}\right]\right] \log 2 , & \text{for } p = \infty . \end{cases}$$

Thus $\sum_p \overline{\lim}_{n\to\infty} h_{p,n}(y) = 2 \log 2 = \rho(y) = \sigma_f(y) + \sigma_\infty(y)$, and

$\overline{\lim}_{n\to\infty} \sum_p h_{p,n}(y) = 3/2 \log 2 = \sigma(y)$.

EXERCISES. 1) Show that $\sigma(y) = 0 \Rightarrow \rho(y) = 0$.

2) Assume that for all $v \in \Sigma(K)$, $\lim_{n\to\infty} h_{v,n}(y)$ exists. Show that $\rho(y) \leq \sigma(y)$.

3) Let $y \in K[[x]]$ and assume that $\rho(y, 1/y) < \infty$,

a) Show that this condition is equivalent to $\sum_{v \in \Sigma_f} \operatorname{Sup}_{n \geq 1} \frac{1}{n} \log^+ |y_n|_v < \infty$ (use the fact that for any $v \in \Sigma_f$, $y^{(v)}$ has no zero $\xi \in \mathbb{C}_v$ satisfying $0 < |\xi|_v < R$, if and only if $r \longmapsto \operatorname{Sup}_n |y_n| r^n$ is a constant function on $]0,R[$),

b) deduce that this condition is satisfied in particular if y is globally bounded,

c) show that $\sigma(y) \leqq \rho(y) < \infty$,

d) deduce that $(\sigma(y^n))_{n\geqq 0}$ is bounded

e) show that if $y(0) \neq 0$, $\sigma(1/y) < \infty$; give upper bounds for $\rho(1/y)$, $\sigma(1/y)$,

f) show that if $y(0) = 0$, then for every z with finite size the composed series $z \circ y$ has again finite size.

4) Consider the series y of § 5.3: assume the finiteness of the set of solutions of the equation $p^k - q^l = m$ (m fixed but arbitrary), and show that, in point of fact, $\sigma(y) = 2^{-N-1}$.

5) Give an example of a series $y \in K[[x]]$ with $R_v(y) = \infty$ for all $v \in \Sigma(K)$.

6) Show that if $\sigma(y) < \infty$, then $R_v(y) < \infty$ for all $v \in \Sigma(K)$ or $y \in K[x]$.

7) Show that the set of diagonals of rational functions is stable under taking derivatives.

Appendix

Calculus of factorials

Following [15]3, we give estimates for the p-adic valuation $v_p((a)_n)$ of the rational number $(a)_n = \prod_{i=0}^{n-1}(a+i)$, for $a \in \mathbb{Q}-(-\mathbb{N})$. We first introduce general notations:

let p be a fixed prime, and let $a \in \mathbb{Q} \cap \mathbf{Z}_p$, i.e. the denominator of a is prime to p.

We define R, Q, and f by the formulae:

$$a = -R(a,p^k) + p^k Q(a,p^k)$$

with $R(a,p^k) \in \mathbb{N}$, $R(a,p^k) < p^k$,

$$f(a,p^k,n) = \left[\frac{n+p^k-1-R(a,p^k)}{p^k}\right].$$

For instance, when $a = 1$, we have $R(1,p^k) = p^k - 1$ and $f(1,p^k,n) = [n/p^k]$. Let us remark that $f(a,p^k,n) - f(1,p^k,n)$ is periodic, with period p^k in n; this leads to the equality

$$(1) \quad f(a,p^k,n) - f(1,p^k,n) = y(\langle n/p^k\rangle - R(a,p^k)/p^k)$$

where $y(x) = \begin{cases} 0 & \text{if } x \leq 0 \\ 1 & \text{if } x > 0 \end{cases}$

and $\langle x\rangle = x-[x]$; we shall also use the notation

$$\{x\} = -x-[-x].$$

We extract from [15][40] a formula for $R(a,p^k)$:

(2) $R(a,p^k)/p^k = \{a\Delta^k\} - a/p^k$ where the integer Δ satisfies the condition:

for some $N \in N^*$, such that $N|a| < p$ and $Na \in \mathbf{Z}$,

$\Delta p \equiv 1 \bmod N$ (in fact $N|a| < p^k$ is enough).

At last we recall the generalization for $(a)_n$, of the classical equality $v_p((1)_n) = \sum_{k=1}^{\infty} [n/p^k]$:

$$(3) \qquad v_p((a)_n) = \sum_{k=1}^{\infty} f(a,p^k,n) .$$

Putting together (1), (2), (3), we find:

LEMMA. The following equality holds:

$$(4) \qquad v_p((a)_n) = \sum_{k=1}^{\infty} [n/p^k] + \#\{ k \text{ such that } \{\Delta^k a\} < (a/p^k + < n/p^k >)\} .$$

REMARK: For $p^k > (a+n)N$, we have $\{\Delta^k a\} \geq 1/N \geq a/p^k + <n/p^k>$, so that the second term at the right-hand side of (4) is bounded by $\frac{\log \text{Max}\ ((a+n)N,0)}{\log p}$.

Chapter II Geometric Differential Equations

§ 1. DEFINITION

1.1 - In this chapter we explain in a precise way what should be the differential equations satisfied by G-functions according to the conjecture stated in the introduction : namely, the "geometric" differential equations over $\bar{\mathbb{Q}}$. These are combinations of factors of Picard-Fuchs equations attached to proper smooth varieties defined over $\bar{\mathbb{Q}}(x)$. We study the stability of this class of differential equations under standard operations and show that their solutions in $\bar{\mathbb{Q}}[[x]]$ form a $\bar{\mathbb{Q}}$-vector space stable under Cauchy and Hadamard products (making use of Hodge theory). We shall show in chapter 5 that such solutions are indeed G-functions.

1.2 - Picard-Fuchs differential equations

Let k be a field of characteristic 0. Let X be a smooth $k(x)$-variety. Its algebraic De Rham cohomology groups $H^i_{DR}(X)$ are $k(x)$-vector space endowed with a canonical connection $\nabla : H^i_{DR}(X) \longrightarrow H^i_{DR}(X) \otimes \Omega^1_X$, called the Gauss-Manin connection, see for instance [38][39] or 2.1 below. Any non-zero vector in this space provides a differential equation with coefficients in $k[x]$.

For X = a completecurve, everything can be made explicit in the following elementary way (after N. Katz): the only interesting group is $H^1_{DR}(X)$, which can be identified with the group of differentials of the second kind on X , modulo the exact ones. Let t be a non-constant function, so that the function field $k(x)(X)$ is a finite extension of $k(x,t)$. Any derivation D of $k(x)$ extends to a derivation D_t on

$k(x)(X)$ by requiring that $D_t(t) = 0$; also the derivation d/dt extends to $k(x)(X)$, and commute with D_t . We let D_t act on differentials by $D_t(f\,dt) = D_t(f)\cdot dt$. The following formulae hold true:

i) $D_t(df) = d(D_t f)$

ii) $res_P(D_t(f\,dt)) = D_t(res_P(f\,dt))$

iii) $(D_t - D_u)(f\,dt) = d(fD_u(t))$.

The first one shows that D_t preserves exactness; by ii), D_t acts as a derivation of $H^1_{DR}(X)$; iii) shows that this action is independent of t , and therefore defines the value at D of a connection ∇ on $H^1_{DR}(X)$: that of Gauss-Manin. Since $H^1_{DR}(X)$ is of finite dimension $2g$, where g = genus of X , we get for any $\omega \in H^1_{DR}(X)$ a relation of the form

$\nabla(d/dx)^{2g}\omega - \sum_{j=0}^{2g-1} \gamma_j \nabla(d/dx)^j\omega = 0$, where $\gamma_j \in k(x)$; multiplying by the common denominator δ of the a_i , we obtain this way an element of $k[x,d/dx]$, namely $\delta(d^{2g}/dx^{2g} - \sum_{j=0}^{2g-1} \gamma_j\, d^j/dx^j)$.

By way of example, take the hypergeometric differential equation associated to $F(a,b,x)$ and discussed in I.4.4: according to W. Messing (see [40]), these are factors of Picard-Fuchs equations.

1.3 - Geometric differential equations

Let $k[x,d/dx]$ be the Weyl algebra over k , sometimes denoted by $A_1(k)$; this is a Noetherian, simple, entire, Euclidean (non-commutative) ring.

DEFINTION. We say that $\Lambda \in A_1(k)$ is a geometric differential equation if Λ is a product of factors (in $A_1(k)$) of some Picard-Fuchs differential equations over k .

In other words, the game of geometric differential equations consists in picking several Picard-Fuchs equations, decomposing them into irreducible factors and then combining some of these factors in arbitrary order. By way of example, take the polylogarithmic equation discussed in I.4.3.. We shall show below that in the definition, it suffices to take Picard-Fuchs differential equations associated to proper smooth varieties over $K(x)$ (using (2.3) and resolution of singularities). Moreover we shall see that the class of geometric differential equations is stable under symmetric powers, and duality (see III ex.2 for definition).

§ 2. GENERALIZATION: MODULES WITH CONNECTION ARISING FROM ALGEBRAIC GEOMETRY

2.1 - The functors $R^i f_*^{DR}$.

Let S be a smooth scheme over a field k of characteristic 0. Let $\underline{MIC}_S$ denote the Abelian category of quasicoherent $\mathcal{O}_S$-modules E with integrable connection $\nabla : E \longrightarrow E \otimes \Omega^1_S$. Let $f : X \longrightarrow S$ be a smooth morphism. One constructs a functor $R^0 f_*^{DR} : \underline{MIC}_X \longrightarrow \underline{MIC}_S$ as follows:

i) as $\mathcal{O}_S$-module, $R^0 f_*^{DR} E = f_*(E^{\nabla | \mathrm{Der}\, X/S})$ where $E^{\nabla|\mathrm{Der}\, X/S}$ denotes the sheaf of germs of horizontal sections of E under the restriction of the connection to the sheaf of germs of relative derivations,

ii) the connection on $f_*(E^{\nabla|\mathrm{Der}\, X/S})$ is defined using the exactness of the sequence of sheaves on X:

$$0 \longrightarrow \mathrm{Der}\, X/S \longrightarrow \mathrm{Der}\, X \xrightarrow{\lambda} f^*\mathrm{Der}\, S \longrightarrow 0 .$$

DEFINITION. For $i \geq 0$, the functor $R^i f_*^{DR} : \underline{MIC}_X \longrightarrow \underline{MIC}_S$ is the ith right derived functor of $R^0 f_*^{DR}$; $R^i f_*^{DR} E$ is the $\mathcal{O}_S$-module $\mathbb{R}^i f_*(\Omega^\bullet_{X/S} \otimes_{\mathcal{O}_X} E)$ endowed with the Gauss-Manin connection.

This makes sense since the categories $\underline{MIC}$ have enough injectives. See [39] or [34], in particular III.4.. We write $H^i_{DR}(X/S)$ for $R^i f_*^{DR}(\mathcal{O}_X, d)$, where d is the universal differential: $\mathcal{O}_X \to \Omega^1_X$. The Picard-Fuchs equations considered above correspond to the case $S = \mathrm{Spec}\, k(x)$.

2.2 - Finiteness properties

We say that a smooth morphism of finite type $f : X \longrightarrow S$

admits a (good) compactification if there exists a diagram

(*)
$$\begin{array}{ccc} X & \hookrightarrow & \bar{X} \\ & \searrow \quad \swarrow & \\ & S & \end{array} \qquad \text{proper smooth } /S$$

such that $D := \bar{X} \setminus X$ is a divisor with relatively normal crossings.

In this situation, let $E \in \mathrm{Ob}\, \underline{MIC}_X$ which extends to an $\mathcal{O}_{\bar{X}}$ -module endowed with a connection with logarithmic singularities along D .

Assume moreover that E is $\mathcal{O}_X$-coherent; then $R^i f_*^{DR} E$ is $\mathcal{O}_S$-coherent, see [21] 6.14. In fact, because characteristic (k) = 0 , any $\mathcal{O}_S$-coherent module with connection is locally free [39] 8.8.

2.3 - The Tannakian categories $\mathcal{G}_S$

DEFINITION. $\mathcal{G}_S$ *is the full sub-category of* MIC_S *which contains the direct summands of* $H^i_{DR}(X/S)$ *for any proper smooth* S-*scheme* X , *and any successive extension of these factors*.

PROPOSITION 1. $\mathcal{G}_S$ *is stable under taking subquotients, tensor product, internal Hom. It contains the objects* $H^i_{DR}(X/S)$ *for any smooth* S-*scheme* X *of finite type which admits a compactification and any* $i \geq 0$.

Proof: we use the fact that the $H^i_{DR}(X/S)$'s for proper smooth X/S are semi-simple objects in $\underline{MIC}_S$. Indeed, by Lefschetz's principle, one may assume that $k = \mathbb{C}$, any by the regularity and comparison theorems [21] 6.13, this semi-simplicity is equivalent to the semi-simplicity of the local system $R^i f_*^{an} \mathbb{C}$, which follows from Hodge-Deligne theory [22] 4.2 (see also [2]).

It follows that in the definition of $\mathcal{G}_S$, one may replace "factor" by "subquotient". On the other hand the objects of $\mathcal{G}_S$ have finite length, according to 2.2. These two facts allow the following reformulation of the definition of $\mathrm{Ob}\mathcal{G}_S$: objects of $\underline{MIC}_S$ whose Jordan-Hölder constituents are constituents of objects $H^i_{DR}(X/S)$ for proper smooth X/S . The stability under taking subquotients follow at once from that description. In order to check the stability under $\otimes$ and duality, it thus suffices, according to that description, to check it on the $H^i_{DR}(X/S)$; this follows from the Künneth formula and the Poincaré duality respectively. Hence G is stable under $\otimes$, duality, and internal Hom.

It remains to prove that for any diagram (*) as in 2.2, $H^i_{DR}(X/S) \in G_S$. By induction on the dimension of X and on the number of components of D , it is enough to prove the following statement: if

$$\begin{array}{ccc} U & \hookrightarrow & Y \\ & \searrow \; S \; \swarrow & \end{array}$$

is a diagram of smooth morphisms such that $Y \setminus U$ is a smooth divisor, then $H^i_{DR}(Y \setminus U/S) \in G_S$ and $H^i_{DR}(Y/S) \in G_S \Rightarrow H^i_{DR}(U/S) \in G_S$. By stability of G_S under extension, this follows from the residue exact sequence

$$\longrightarrow H^{i-2}_{DR}(Y \setminus U/S) \longrightarrow H^i_{DR}(Y/S) \longrightarrow H^i(U/S) \xrightarrow{Res} H^{i-1}(Y \setminus U/S) \longrightarrow \dots ,$$

which arises from the contravariance of De Rham homology with respect to open immersions (see [34] III.4 Note p.74). The recursion goes through because at each step the morphism $Y \setminus U \to S$ admits an obvious compactification.

□

COROLLARY. *Assume that* S *has a rational point* s *over* k . *Then* G_S *is equivalent to the category of* k-*rational representations of an affine group scheme over* k .

Proof: we use the theory of Tannakian categories, see e.g. [22] II. The previous proposition shows that G_S is an Abelian k-linear rigid tensor category. The stalk at s furnishes an exact functor $G_S \longrightarrow$ k-vector spaces. Applying to Hom the fact that horizontal sections are determined by any stalk, we get the faithfulness of this functor. Summarizing, we obtain this way a neutral Tannakian category, and the fundamental result of the theory gives the expected conclusion.

□

REMARK. For a comparison between the Galois group of G_S and a motivic Galois group, we refer the reader to [2] or [3].

Let $f : X \longrightarrow S$ be a smooth morphism of finite type which admits a compactification.

PROPOSITION 2. The functor $R^i f_*^{DR} : \underline{MIC}_X \longrightarrow \underline{MIC}_S$ carries $\mathcal{G}_X$ into $\mathcal{G}_S$.

Proof: let $Z \longrightarrow X$ be a proper smooth morphism. By the same argument as in I, making use of the Lefschetz-Deligne theorem, the Leray spectral sequence $R^i f_*^{DR}(H^j_{DR}(Z/X)) \Rightarrow H^{i+j}_{DR}(Z/S)$ degenerates. By the previous proposition, this shows that for any direct summand E of $H^j_{DR}(Z/X)$, $R^i f_*^{DR} E \in \mathrm{Ob}\, \mathcal{G}_S$ (one uses here that $R^0 f_*^{DR}$ commutes with finite direct sums, and so does $R^i f_*^{DR}$ henceforth). One extends this to the case of extensions of such E's by using the fact that $R^i f_*^{DR}$ is a cohomological functor by definition: one writes the associated long exact sequence.

□

§ 3. SOLUTIONS OF GEOMETRIC DIFFERENTIAL EQUATIONS; ALGEBRAIC STRUCTURE

We turn back to the situation of § 1.

THEOREM. The set of solutions in $k[[x]]$ of geometric differential equations is a k-sub-vector space of $k[[x]]$ stable under usual (= Cauchy) and Hadamard products.

Note the following immediate consequence:

COROLLARY. If $\Sigma y_n x^n$ satisfies a geometric differential equation, so does $\Sigma y_n^N x^n$ for any $N \geq 0$.

Proof: let y_1, y_2 two such solutions, in the "kernel" of elements Λ_1 , resp. Λ_2 of $A_1(k)$. Making use of Ore's localizability condition in $A_1(k)$, one may find a common multiple for Λ_1 and Λ_2 : $\Lambda = \Gamma_1\Lambda_1 = \Gamma_2\Lambda_2$; hence Λ annihilates both y_1 and y_2 , and consequently any linear combination of the two. The problem is to find such a Λ which is a geometric differential equation. This can be done in the following manner: Λ_i is given by an object E_i in $\mathcal{G}_S$ for $S = \mathrm{Spec}\, k(x)$ and an element e_i inside E_i . For Λ , take

$E = E_1 \oplus E_2 \in \mathrm{Ob}\, G_S$, together with $e = (e_1,e_2)$. The solutions y_i of Λ_i correspond to $\theta_i(e_i)$ for some horizontal $\theta_i : E_i \longrightarrow k((x))$. Taking $\theta = \lambda\theta_1 \oplus \mu\theta_2$, one can see that the solutions of geometric differential equations form indeed a k-vector space. The product y_1y_2 is a solution of the symmetric square $\Lambda^{\odot 2}$, which corresponds to a quotient of $E \otimes E$. Because such a quotient is an object of G_S (prop.1), $\Lambda^{\odot 2}$ is indeed a geometric differential equation.

Now consider $y_i(x_i)$ as functions of independent indeterminates x_1,x_2 , and let us write $\theta_i : E_i \longrightarrow k((x_i))$. We may consider the exterior tensor product $E_1 \boxtimes E_2$ as an object of $\underline{MIC}_U$ for U a dense open subset of $W = \mathrm{Spec}\; k(x)[x_1,x_2]/_{(x_1x_2-x)} \approx \mathbb{G}_m|_{k(x)}$. The morphism $f : U \longrightarrow S = \mathrm{Spec}\; k(x)$ admits obviously a compactification.

It follows from propositions 1 and 2 that $R^1f_*^{DR}(E_1 \boxtimes E_2) \in G_S$. On the other side, § 3.3 of chapter I tells us that $y_1 * y_2$ is a solution of the differential equation obtained from $(R^1f_*^{DR}(E_1 \boxtimes E_2)$, class of $y_1(x_1)y_2(x_2)\frac{dx_2}{x_2})$. Hence $y_1 * y_2$ satisfies a geometric differential equation.

□

EXERCISES. 1) Using some results from Hodge theory [22] or [3], show that any rank-one module with connection arising from algebraic geometry becomes trivial over an étale finite covering of the base.

2) Describe $R^0f_*^{DR}$ when f is a finite morphism onto $\mathrm{Spec}\; k(x)$; in particular, give an interpretation of the Picard-Fuchs equations associated to $H^0_{DR}(f)$. Show directly that $H^0_{DR}(f)$ is a semi-simple object in $\underline{MIC}_{k(x)}$.

Part Two

G-Functions and Differential Equations

Chapter III Fuchsian Differential Systems: Formal Theory

§ 1. LOGARITHMIC SINGULARITIES

1.1 - Canonical form of solutions

In this chapter, k denotes an arbitrary field of characteristic 0. We write $\partial = x\, d/dx$ for the derivation of $M_\mu(k((x)))$ given by $\partial(\Sigma Y_n x^n) = \Sigma n Y_n x^n$.

Let $G \in M_\mu(k[[x]])$ such that the eigenvalues of $G(0)$ belong to k (i.e. the characteristic polynomial of $G(0)$ splits in k). One considers the following differential system, with logarithmic (= regular = Fuchsian) singularity at 0:

(1) $\partial X = GX$.

The classical theory (see [31] ch. 14 for instance) tells us how to transform (1) into the following equivalent system:

(2) $\partial(Y^{-1}X) = CY^{-1}X$, that is, formally: $X = Yx^C = Y \exp(C \log x)$,

(3) $\partial Y = GY - YC$, with $Y \in GL_\mu(k((x)))$ and $C \in M_\mu(k)$.

If in addition to the previous assumptions, one supposes that

(4) none of the differences between the eigenvalues of $G(0)$ is a non-zero integer,

one can then choose $C = G(0)$; there is a unique solution $Y_G \in GL_\mu(k[[x]])$ of (3) compatible with this choice such that $Y_G(0) = I$. In analogy with the analytic case, we call Y the <u>uniform part</u> of X , and Y_G (under (4)) the <u>normalized uniform part</u>. Making use of the formula

(5) $\partial Y^{-1} = -Y^{-1}\, \partial Y\, Y^{-1}$, it is straightforward to check that if (Y', C') is another solution of (3), the matrix $V := (Y')^{-1} Y \in GL_\mu(k((x)))$ then satisfies the differential system

(6) $\partial V = C'V - VC$.

1.2 - Analysis of the system (6)

We consider equation (6) for itself. Let us write $V = \Sigma V_n x^n$, and put $U(C',C)$ for the linear endomorphism $A \longmapsto C'A - AC$ of $M_\mu(k)$.

Equating coefficients, one sees that (6) is equivalent to the list of equations:

$$(7)\quad U(C',C)V_n = nV_n .$$

This means that the non-zero coefficients of V are eigenvectors of $U(C',C)$ with distinct eigenvalues. It follows that $V \in M_\mu(k[x,1/x])$.

On the other hand, it is well-known that the eigenvalues of $U(C',C)$ are the differences between the eigenvalues of C' and those of C; it follows that if C and C' are conjugated and satisfy (4) (with C instead of $G(0)$), then $V \in M_\mu(k)$.

1.3 - Duality

The dual system of (1) is

$$(1)^*\quad \partial X^* = -{}^tG\, X^* .$$

One can pass from solutions of (1) to solutions of (1)* by setting $X^* = {}^tX^{-1}$.

Moreover, if (Y,C) satisfies (3), then $({}^tY^{-1}, -{}^tC)$ satisfies the corresponding system (3)* associated to (1)*.

1.4 - Writing out the uniform part

In order to express the action of the "higher derivation" $x^m d^m/dx^m$ on the solutions of (1), one defines recursively a sequence of matrices $G_{[m]} \in M_\mu(k[[x]])$:

$$(8)\quad G_{[0]} = I,\ G_{[m+1]} = \partial G_{[m]} + G_{[m]}(G-mI) .$$

With that notation, one finds the equation $x^m d^m/dx^m X = G_{[m]}X$ whenever X formally satisfies (1). Moreover one computes readily:

$$(9)\quad \frac{G_{[m]}(0)}{m!} = \binom{G(0)}{m} = \frac{1}{m!}\left[G(0)-I)\ldots(G(0)-(m-1)I)\right] .$$

Starting from the elementary calculations

$$G_{[m+1]}Y = \partial G_{[m]}Y + G_{[m]}(G-mI)Y = \partial(G_{[m]}Y) + G_{[m]}Y(C-mI)$$

and $\binom{C+n}{m}((n-m)I+C) = (m+1)\binom{C+n}{m+1}$, one finds inductively the formula

(10) $\frac{G_{[m]}}{m!} Y = \Sigma Y_n \binom{C+n}{m} x^n$.

Equating coefficients, this means that

(11) $Y_n \binom{C+n}{m} = \sum_{l \leq n} \frac{1}{m!} G_{[m]l} Y_{n-l}$.

For $m = 1$, making use of the (previously defined) operator U , one translates (11) into:

(12) $U(C+nI, G(0))Y_n + \sum_{l<n} G_l Y_{n-l} = 0$.

Let us now assume that condition (4) holds. It follows that $U(G(0)+nI, G(0))$ is invertible for $n \geq 1$; its inverse has the following form: $\prod_{i,j<\mu} (\alpha_i - \alpha_j + n)^{-1} \cdot \mathrm{Adj}\, U(G(0)+nI, G(0))$, where α_i, $i = 0,\ldots,\mu-1$ runs over the family of eigenvalues of $G(0)$. This gives an induction formula for the nth coefficient of the normalized uniform part:

(13) $Y_{G,n} = \left(\prod_{i,j<\mu} (\alpha_i - \alpha_j + n)^{-1}\right) \cdot \left(\mathrm{Adj}\, U(G(0)+nI, G(0))\right) \cdot \sum_{m=0}^{n-1} G_{n-m} Y_{G,m}$.

This way, we get a proof of the equivalence between (1) and {(2),(3)} under (4). One may reduce the general case to this special case by means of some shearing transformations, see below.

§ 2. THE LANGUAGE OF ∂-MODULES

2.1 - We now interpret § 1 in the more canonical and somewhat "dual" language of ∂-modules, which will provide a suitable way to express our results.

Let A an entire commutative unitary k-algebra containing $k[x]$; we assume that the derivation ∂ on $k[x]$ extends to A , and that $\ker\partial = k$. By a ∂-module over A , we simply mean a unitary module M over the non-commutative ring $A[\partial]$, such that the induced A-module is free of finite type.

If $A \subset k[[x]]$, the action of ∂ over this induced A-module defines the usual notion of a connection with logarithmic singularity along 0.

Let us endow M with an A-basis $\{m_i\}_{i=0}^{\mu-1}$, and let us write

(14) $\partial m_i = \sum_{j=0}^{\mu-1} {}_{ij}G\, m_j$ for $i = 0,\ldots,\mu-1$.

Though ∂ is not A-linear, we shall say by abuse of language

that G <u>represents</u> ∂ in the basis m_j .
Now let K be an over-ring of A , over which the derivation ∂ extends, such that $\ker\partial = k$. We denote by $S(M,K)$ the k-vector space formed by the <u>solutions of</u> M <u>in</u> K , i.e. by the $A[\partial]$-morphisms from M to K . One has $\dim_k S(M,K) \leq \mu$, and in case of equality, one says that M is <u>solvable</u> in K . In this case the matrix $X \in GL_\mu(K)$ formed by the components of a k-basis of $S(M,K)$ in the A-dual basis $\{m_i^*\}$ of $\{m_i\}$ satisfy the differential system (1); the components of a column of X are the values of $\theta \in S(M,K)$ at $m_0,\dots,m_{\mu-1}$. Conversely, system (1) describes solutions of a unique structure of ∂-module over $k[[x]]^\mu$ endowed with its canonical basis. The dual ∂-module M^* of M is $\mathrm{Hom}_A(M,A)$ with "connection" represented by $-{}^tG$ in the dual basis $\{m_i^*\}$. This is compatible with 1.3.

2.2 - <u>Canonical form</u>
By conjugating system (1), that is to say X and G , by a matrix in $GL_\mu(k)$, one can reduce the matrix C appearing in (2), (3) into its canonical Jordan form. Let us denote by $[\alpha]_A$ the twist by $\alpha \in k$: this means the ∂-module structure on A defined by $\partial 1 = \alpha$. One also denotes by L_A^ν , for $\nu \in \mathbb{N}^x$, the ∂-module structure on A^ν where ∂ is represented in the canonical basis by the Jordan block of size ν .
Now assume that $A \subseteq k[[x]]$. Making use of the dictionary 2.1, the equivalence between (1) and (2),(3) can then be stated as follows: $M \otimes_A k((x))$ is isomorphic to one of the standard ∂-modules $\bigoplus_i [\alpha_i]_{k((x))} \otimes L_{k((x))}^{\nu_i}$; where α_i runs through the eigenvalues of C . Moreover the ∂-modules $L_{k((x))}^{\nu_i}$ are indecomposable, and the above standard form for $M \otimes_A k((x))$ is unique up to translation of the α_i mod $\mathbb{Z}$, see [48].

2.3 - <u>Exponents</u>
For $A \subseteq k[[x]]$, the eigenvalues of $G(0)$ (multiplicities taken into account) depend only on the ∂-module M and not of the chosen basis which defines G : we shall call these eigenvalues are the <u>exponents</u> of M . When they satisfy condition (4) above, we shall say the M is <u>normalized</u>.

2.4 - Morphisms

Let A be as in 2.1. Let M, M' be two ∂-modules over A, endowed with bases $\{m_j\}$ and $\{m'_j\}$ respectively. Let H be the matrix of a morphism (of $A[\partial]$-modules) from M to M' in those bases, i.e.:

(15) $\text{image}\ (m_i) = \sum_j {}_{ij}H\ m'_j$.

Let G, resp. G' be the matrices which represent ∂ in M (resp. M') with recourse to $\{m_i\}$ resp. $\{m'_i\}$. One readily check that H satisfies the following differential system:

(16) $\partial H = GH - HG'$.

One can pass from solutions of (1)': $\partial X' = G'X'$ to solutions of (1) by setting

(17) $X = HX'$.

Let us now assume that H is invertible in $GL_\mu(A)$; this means that H represents an isomorphism of ∂-modules. Let $V := (HY')^{-1}Y$ be the matrix relating the uniform parts of solutions of (1) and (1)'. Using (5) and (3), one finds

$$\begin{aligned}\partial V &= -Y'^{-1}H^{-1}\partial H\, H^{-1}Y - Y'^{-1}\partial Y'.Y'^{-1}H^{-1}Y + Y'^{-1}H^{-1}\partial Y\\ &= -Y'^{-1}H^{-1}GY + Y'^{-1}G'H^{-1}Y - Y'^{-1}(G'Y' - Y'G')Y'^{-1}H^{-1}Y + Y'^{-1}H^{-1}(GY - YC)\\ &= C'V - VC\ .\end{aligned}$$

This is equation (6), which was analyzed in 1.2. In particular, one gets that if condition (4) is fulfilled, and if $H(0)$ is invertible, then:

(18) $Y_G = H\ Y_{H[G]}H(0)^{-1}$. Here we have set:

(19) $H[G] := H^{-1}GH - H^{-1}\cdot\partial H$, to denote G' .

In particular if $M = M'$, these formulae give the behaviour of representative matrices through any change of basis. Roughly speaking, "∂-module" is an intermediate notion between differential systems and connections: we allow changes of basis, but no change of variable . In the next paragraph, we shall study two special cases of change of basis.

§ 3. SPECIAL CHANGES OF BASIS

3.1 - Shearing

This is a tool for modifying the exponents by integers in order to reach the normalized situation (condition (4)), see [31] ibid. Let G be as in 1.1 and consider the set $\{\alpha_0,\ldots,\alpha_\nu\}$ of the eigenvalues of $G(0)$. Fix a set of integers $\{n_0,\ldots,n_\nu\}$ with the same cardinality as $\{\alpha_0,\ldots,\alpha_{\mu-1}\}$.

Let S the mätrix of a new basis of k^μ adapted to the decomposition $\bigoplus_{i=0}^{\nu} \ker(G(0)-\alpha_i I)^{\text{mult }\alpha_i}$ into characteristic subspaces. Thus $S^{-1}G(0)S$ is formed by diagonal blocks, each of which has only one eigenvalue. One can moreover assume that this is triangular (e.g. $SG(0)S^{-1}$ has Jordan's normal form). Now let Δ be the diagonal matrix

$$\begin{pmatrix} x^{n_0} & & & & & & \bigcirc \\ & x^{n_0} & & & & & \\ & & \ddots & & & & \\ & & & x^{n_i} & & & \\ & & & & x^{n_i} & & \\ & & & & & \ddots & \\ \bigcirc & & & & & & x^{n_\nu} \end{pmatrix} \begin{matrix} \Big\} \text{ mult } \alpha_0 \\ \Big\} \\ \vdots \end{matrix}$$

, and put $H^{sh} = S\,x^{-\Delta}$.

By formula (19) we get $H^{sh}[G] = +\,\Delta + x^{+\Delta}S^{-1}GS\,x^{-\Delta}$, with coefficients ${}_{ij}(H^{sh}[G]) = {}_{ij}\Delta + x^{\Delta_i-\Delta_j}\,{}_{ij}(S^{-1}GS)$. It is thus clear that:

(20) $H^{sh} \in GL_\mu(k[x,1/x])$

(21) $H^{sh}[G] \in M_\mu(k[[x]])$.

Furthermore, by looking at the diagonal terms of $H^{sh}[G](0)$, one sees that

(22) the set of eigenvalues of $H^{sh}[G](0)$ is $\{\alpha_i+n_i\}$.

3.2 - Reduction to a differential equation

We assume in this paragraph that (1) is solvable in $GL_\mu(k((x)))$; sometimes one says that 0 is a cosingularity of this differential system (see e.g. [59] p.7; this is a property of the associated ∂-module over $k((x))$). It is clear that in this case there exists some positive integer l such that $-l \leq \mathrm{ord}_0 f \leq l$ for any f in the k-vector space spanned by the entries of a solution X of (1) in $GL_\mu(k((x)))$. Let l_j, $j=0,\ldots,\mu-1$, be integers such that $\mathrm{Max}(l_j-l_i, l_i-l_j)>2l$ for $i\neq j$. The following fact follows easily from these choices: for any $(\alpha_i)_{i=0}^{\mu-1} \in (k\setminus\mathbb{Q})^\mu$, the series $y_j = \sum_{i=0}^{\mu-1} \alpha_i\, x^{l_j}\, {}_{ij}X$, $j = 0,\ldots,\mu-1$, are linearly independent over k.

Set ${}_{nj}S := \sum_{m=0}^{n} \sum_{i=0}^{\mu-1} \binom{n}{m} \alpha_i x^{m-n}\left(\frac{d^m x^{l_i}}{dx^m}\right)\cdot\left({}_{ij}G_{[n-m]}\right)$, and

$S = ({}_{ij}S)_{i,j=0}^{\mu-1}$. Making use of Leibnitz rule, one checks easily that the rows of SX are the successive derivatives of $(y_0,\ldots,y_{\mu-1})$. Because of the linear independence of these functions, S lies in $GL_\mu(k((x)))$; putting $H^{eq} = S^{-1}$, one sees that system (1) is equivalent to the differential equation

$$(23)\quad \frac{d^\mu}{dx^\mu}\, y = \sum_{j=0}^{\mu-1} {}_{\mu-1,j}\!\left(\frac{H^{eq}[G]}{x}\right) \frac{d^j y}{dx^j}\,.$$

More precisely, if $y_0,\ldots,y_{\mu-1}$ are linearly independent solutions of (23) in $k((x))$, one get a solution of (1) by setting

$$X = H^{eq}\left(\frac{d^i y_j}{dx^i}\right)_{i,j=0,\ldots,\mu-1}.$$

The matrix $H^{eq}[G]$ looks like this:

$$\begin{pmatrix} 0 & x & & \bigcirc \\ & & \ddots & \\ \bigcirc & & & x \\ * & * & \cdots & * \end{pmatrix}.$$

REMARK 1. Let $(*)\ \frac{d^\mu}{dx^\mu}\, y = \sum_{j=0}^{\mu-1} \gamma_j \frac{d^j y}{dx^j}$ denote a differential equation with coefficients $\gamma_j \in k((x))$. It is easy to translate it into $\partial^\mu y = \sum_{j=0}^{\mu-1} g_j \partial^j y$, or else (1) $\partial X = GX$ by setting

$$G = \begin{pmatrix} 0 & 1 & & \bigcirc \\ & & \ddots & \\ \bigcirc & & & 1 \\ g_0 & g_1 & \cdots & g_{\mu-1} \end{pmatrix}$$ as usual.

By Fuchs' theory, or elementary calculation, one obtains that $G \in M_\mu(k[[x]])$ iff $\mathrm{ord}_0 \gamma_j \geq j-\mu$. The characteristic polynomial of G is called the indicial polynomial of (*). Note that if 0 were an ordinary point for (*) (i.e. $\gamma_j \in k[[x]]$, it remains no longer an ordinary point for the associated system (1); indeed, the exponents are $0,1,\ldots,\mu-1$.

REMARK 2. We take this opportunity to make a general remark about components of column-solutions of (1). By 2.1, we know that they have the shape $\theta(m_i)$ for some $\theta \in \mathrm{Sol}(M,K)$, and m_i running over a basis of M (the associated ∂-module). Since $\mathrm{rk}\, M \leq \mu$, $\partial^\mu m_i$ is linear combination of $\partial^j m_i$, $j = 0,\ldots,\mu-1$ with coefficients in the base ring A . Since θ is $A[\partial]$-linear, $\partial^\mu\theta(m_i)$ is the same combination of the $\partial^j\theta(m_i)$. Hence any component of a solution of a differential system of order one satisfies a scalar differential equation of order $\leq \mu$ with coefficients in A .

§ 4. BLOW UP

This construction of Bombieri-Sperber [8] provides a very simple "model" for differential equations with given exponents, according to the regularity. We follow the presentation in loc.cit.

Let $\Lambda \in k((x))[d/dx]$ be a differential operator of order μ . We denote by Λ_m the unique element in the ideal $k((x))[d/dx]\Lambda$ such that the degree of $\Lambda_m - \frac{1}{m!} d^m/dx^m$ is at most $\mu-1$, and we write $\Lambda_m = \frac{1}{m!}\frac{d^m}{dx^m} - \Sigma\gamma_{m,j}\frac{1}{m!} d^j/dx^j$. The coefficients $\gamma_{m,j}$ satisfy the recurrence

(24) $\gamma_{m+1,j} = \frac{1}{m+1}(d/dx\, \gamma_{m,j} + j\gamma_{m,j-1} + \mu\gamma_{m,\mu-1}\gamma_j)$.

Let us write $\gamma_{\mu,j} = \frac{\lambda_j}{x^{\delta_j}}$ + higher terms, $\lambda_j \neq 0$, and let us define: $\rho := \max_{0\leq j<\mu}(\delta_j/_{\mu-j},1)$, so that 0 is a logarithmic singularity if and only if $\rho = 1$, and let us put

$$J := \{0 \leq j < \mu : \delta_j = \rho(\mu-j)\} .$$

The recursion (24) implies that $\gamma_{m,j} = \frac{\lambda_{m,j}}{x^{\rho(m-j)}} + \text{h.o.t.}$, where of course $\lambda_{m,j} = 0$ if $\rho(m-j)$ is not an integer and $\lambda_{\mu,j} = \lambda_j$ if $j \in J$. One defines the blow up of Λ to be the constant coefficient differential operator

$$(25)\quad \tilde{\Lambda} = \frac{1}{m!}\frac{d^\mu}{dx^\mu} - \sum_{j\in J}\lambda_j \frac{1}{j!}\frac{d^j}{dx^j} \quad \text{if } \rho > 1 ,$$

and the differential operator of Euler type

$$(26)\quad \tilde{\Lambda} = \frac{1}{\mu!}\frac{d^\mu}{dx^\mu} - \sum_{j\in J}\frac{\partial_j}{(1+x)^{\mu-j}}\frac{1}{j!}\frac{d^j}{dx^j} \quad \text{if } \rho = 1 .$$

Note that in the latter case, the exponents of X at -1 are the exponents of Λ at 0.

LEMMA. <u>The formal power series</u> $y_j = x^j + \lambda_{\mu,j}x^\mu + \lambda_{\mu+1,j}x^{\mu+1} + \ldots$ <u>are annihilated by</u> $\tilde{\Lambda}$, for $j = 0,\ldots,\mu-1$.

Proof: let $z_j(t+x) \in k((t))[[x]]$, for $j = 0,\ldots,\mu-1$, be defined by the formula $z_j(t+x) = \sum_{m\geq 0}\gamma_{m,j}(t)x^m$. It is clear that $\frac{1}{h!}\frac{d^h}{dx^h} z_j(t+x)\Big|_{x=0} = \begin{cases}1 & \text{if } h = j\\ 0 & \text{if } h \neq j\end{cases}$ for $h = 0,1,\ldots,\mu-1$, and that $\frac{1}{\mu!}\frac{d^\mu}{dx^\mu} z_j(t+x)\Big|_{x=0} = \gamma_{\mu,j}(t)$. Hence $\Lambda z_j = 0$. Since $\gamma_{m,j}(t) = \frac{\lambda_{m,j}}{t^{\rho(m-j)}} + \text{h.o.t.}$, we obtain $t^{-\rho j}z_j(t+t^\rho x) = y_j(x) + t^{1/b}\psi_j(t^{1/b},x)$ where the natural integer b is such that $b\rho$ is an integer, and where $\psi_j(\tau,x) \in k[[\tau,x]]$. It follows that y_j is obtained from $t^{-\rho j}z_j(t+t^\rho x)$ by specializing t to 0.

Let $\tilde{\Lambda}_t \in k((t))[[x]][d/dx]$ be defined by

$$\tilde{\Lambda}_t = \frac{1}{\mu!}\frac{d^\mu}{dx^\mu} - \sum_{h=0}^{\mu-1}\gamma_{\mu,h}(t+t^\rho x)t^{(\mu-h)\rho}\frac{1}{h!}\frac{d^h}{dx^h} ,\ \text{so that}$$

$\tilde{\Lambda}_t(t^{-\rho j}z_j(t+t^\rho x)) = 0$. The lemma follows by noting that $\tilde{\Lambda}_t$ specializes to $\tilde{\Lambda}$ when t specializes to 0.

E. Bombieri and S. Sperber interpret this proof as performing a blowing up transformation $(x,z) \longrightarrow (\frac{x-t}{t^\rho}, t^{-\rho j}z)$ for $t \longrightarrow 0$ on the graph of the mapping $x \longrightarrow z_j(x)$ and obtaining the differential operation associated to the blown up graph.

§ 5. FORMAL FROBENIUS AND CHRISTOL FUNCTORS

5.1 - The Φ functor

We turn back to the general assumptions of 2.1. Let p be a fixed prime number. For any automorphism φ of k, we consider the so-called Frobenius endomorphism of $k[x]$, denoted by $\phi : f \mapsto f^{\varphi}(x^p)$, where φ applies to the coefficients of f. We have the identity:

(27) $\partial\phi(f) = p\phi(\partial f)$.

We assume that ϕ extends to an injective endomorphism of A such that (27) carries over.

For $G \in M_\mu(A)$, we put $G^\phi := G^\varphi(x^p)$, with the same meaning as above.

Let $M, \{m_i\}, G$ be as in 2.1. We denote by M^ϕ the ∂-module over A such that:

i) as an A-module, $M^\phi = M$

ii) the action of ∂ in the basis $\{m_i\}$ of M is represented (in the sense of 2.1) by pG^ϕ .

If K is any over-ring of A on which ϕ extends such that (27) remains true, any solution X of (1) in $GL_\mu(K)$ satisfies: $(1)^\phi$ $\partial X^\phi = pG^\phi X^\phi$. Similarly, we have $(Y_G)^\phi = Y_{pG^\phi}$ in the normalized case. It is readily checked that ϕ defines a functor on the category of $(M, \{m_i\})$, thus on the category of ∂-modules up to isomorphism; ϕ is called the Frobenius functor. It commutes with direct sums and tensor products. For the twist $[\alpha]$ (see 2.2), we have $[\alpha]^\phi = [p\alpha]$.

5.2 - Instead of 0, one can choose another "center" $a \in k$, and replace $x \mapsto x^p$ by $(x-a) \mapsto (x-a)^p$; this way one get a functor ϕ_a acting on (isomorphy classes of) $(x-a)d/dx$-modules.

5.3 - The Ξ functor

We shall be interested for later purposes in inverting ϕ .

However, one cannot expect in our general situation that ϕ is quasi-invertible; so we shall content ourselves here in constructing a functor Ξ which inverts ϕ up to some twists $[i/p]$. Let us assume in addition to the previous hypotheses:

(28) A is a free $\phi(A)$-module generated by $1, x, \ldots, x^{p-1}$.

This is the case, for instance, if $A = k[x]$ or $k(x)$. More generally, if $A \subseteq k((x))$, this is the case if and only if A is stable under the additive maps

$$\psi_i : \Sigma f_n x^n \longmapsto \Sigma f_{i+np}^{\varphi^{-1}} x^n \quad i = 0, \ldots, p-1 ,$$

since the following identity holds in $k((x))$:

$$(29) \quad f = \sum_{i=0}^{p-1} x^i \phi \psi_i f .$$

In general, take (29) as the definition of the ψ_i ; it is a correct one since ϕ is assumed to be injective. The identity (29) (and definition of ψ_i) extends readily to the case of matrices, and it follows from (27) that

$$(30) \quad \partial \psi_i G = \frac{1}{p} \psi_i \partial G - \frac{i}{p} \psi_i G , \text{ for any } G \in M_\mu(A) .$$

Now for any such G ,we follow [11]3 and set:

$$(31) \quad G^{\Xi} := \begin{pmatrix} \psi_0 G & x\psi_{p-1} G & \cdots & x\psi_1 G \\ \psi_1 G & \psi_0 G & & x\psi_2 G \\ \vdots & & \ddots & \\ \psi_{p-1} G & & & \psi_0 G \end{pmatrix} \in M_{p\mu}(A) ,$$

$$\text{and} \quad J := \begin{pmatrix} 0 & 0 & & \\ 0 & 1/p\, I_\mu & & 0 \\ & & \ddots & \\ 0 & & & \frac{p-1}{p} I_\mu \end{pmatrix} \in M_{p\mu}(k) .$$

Starting from (29), applied to GF :

$GF = \sum_{i=0}^{p-1} x^i (\psi_i G)^\phi \sum_{j=0}^{p-1} x^j (\psi_j F)^\phi$, one arrives to the formula

$$\psi_i(GF) = \sum_{j=0}^{i} (\psi_j G)(\psi_{i-j} F) + x \sum_{j=i+1}^{p-1} (\psi_j G)(\psi_{i-j+p} F) ,$$

which implies in turn that

(32) $(GF)^{\Xi} = G^{\Xi} F^{\Xi}$.

On the other hand, it follows from (30) that

(33) $\partial(G^{\Xi}) = \frac{1}{p}(\partial G)^{\Xi} - JG^{\Xi} + G^{\Xi}J$.

Let $M, \{m_i\}, G$ be as in 2.1. We denote by M^{Ξ} the ∂-module over A such that:

i) as an A-module, $M^{\Xi} = M \oplus \ldots \oplus M$ (p factors)

ii) the action of ∂ in the basis $\{m_i\}$ of M is represented by $\frac{1}{p} G^{\Xi} - J$.

It is straightforward to deduce from (32) and (33) that Ξ defines a functor on the category of $(M, \{m_i\})$, thus on the category of ∂-modules up to isomorphism. Let us call Ξ the <u>Christol functor</u>. The following statement is due to G. Christol [11].

PROPOSITION. <u>Let</u> A <u>satisfy the above hypotheses, and let</u> M <u>be a</u> ∂<u>-module over</u> A . <u>Then</u> $(M^{\phi})^{\Xi} \simeq \bigoplus_{i=0}^{p-1} ([i/p] \otimes M)$.

Proof: this follows from the fact that

$$(G^{\phi})^{\Xi} = \begin{pmatrix} G & & & \bigcirc \\ & G - \frac{\mu}{p} I & & \\ & & \ddots & \\ \bigcirc & & & G - \frac{p-1}{p} I_{\mu} \end{pmatrix} .$$

□

EXERCISES. 1) By means of Gantmacher's counterexample $G = \begin{pmatrix} 0 & x \\ 0 & 1 \end{pmatrix}$, test the statements of § 1 without (4) .

2) Describe the ∂-module $\mathrm{Hom}(M,N)$ in terms of matrices representing M and N .

3) (O. Gabber) Assume that M admits an A-basis $\{m_i\}_{i=0}^{\mu-1}$ such that $m_{i+1} = \partial m_i$; one says that m_0 is a cyclic vector. Show that there is a relation $(\partial^{\mu} + \sum_{i=0}^{\mu-1} a_i \partial^i) m_0 = 0$. Show that the $(\mu-1)^{\underline{th}}$ element $m^*_{\mu-1}$ of the dual base is a

cyclic vector for M^*, and satisfies the dual differential equation $((-\partial)^{\mu} + \sum_{i=0}^{\mu-1} (-\partial)^{i} a_i) m^*_{\mu-1} = 0$.

4) (G. Christol) under the assumptions of 5.3, let $H \in GL_{\mu}(A)$. Show that $H^{\Xi} \in GL_{p\mu}(A)$, at least for $A \subseteq k((x))$. Hint: introduce the matrix $(\zeta^{ij} x^{i} I_{\mu})_{i,j=0}^{p-1}$.

5) Using 2.2, prove that, for $A = k((x)))$ $M^{\phi} \simeq \eta^{\phi} \Rightarrow \bigoplus_{i=0}^{p-1} [i/p] \otimes M \simeq \bigoplus_{i=0}^{p-1} [i/p] \otimes \eta$ (without making use of Ξ).

APPENDIX: A ZERO ESTIMATE

Following a method of G.V. Chudnovsky [17] we sketch the proof of an explicit upper bound for the order at one point of certain polynomial combinations of solutions of a Fuchsian differential system over $A = k(x)$. This result sharpens Shidlovski's lemma (see [55]) in the Fuchsian case, and can be extended in a non-effective way to the general case according to Bertrand-Beukers [6].[1]

Let us assume that $k = \mathbb{C}$ and let us consider the following differential system:

(I) $d/_{dx}\, X = \Gamma X$ with $\Gamma \in M_{\mu}(k(x))$.

We make the following assumptions:

i) (I) is in the Fuchsian class, i.e. at any pole s of Γ in $\mathbb{P}^1_k$ there exist matrices $Y_s \in M_{\mu}(k[[x-s]])$ of rank μ, $C_s \in M_{\mu}(k)$, such that $Y_s(x-s)^{C_s}$ satisfies (I); here we set $x - \infty := 1/x$,

ii) there exists a solution $Y = {}^t(y_0,\ldots,y_{\mu-1})$ of (I) in $k[[x]]^{\mu}$.

Let p be an element of $k[x,x_0,\ldots,x_{\mu-1}]$; we assume that $\deg_x p = N$ and that p is homogeneous of degree n in $x_0,\ldots,x_{\mu-1}$. We shall study the order at 0 of $r := p(x,y_0,\ldots,y_{\mu-1}) \in k[[x]]$. For $s \in \mathrm{Sing}\Gamma$ (the set of poles of Γ in $\mathbb{P}^1_k$), and for C_s as above, we denote by ε_s the minimum of the real parts of the eigenvalues of C_s. At last, let μ_r denote the dimension of the $k(x)$-vector space generated by the successive derivatives of r , so that $\mu_r \leqq \binom{\mu+n-1}{n}$.

THEOREM. Either $r = 0$, or

$$(34) \quad \mathrm{ord}_0 r \leqq N\mu_r - n\mu_r \sum_{\mathrm{sing}\Gamma} \varepsilon_s + (|\mathrm{sing}\Gamma|-2)\frac{\mu_r(\mu_r-1)}{2} .$$

Replacing (I) by its $n^{\underline{th}}$ symmetric power, it is plain to see that it suffices to prove the theorem for $n = 1$. The proof relies on the following lemma, which belongs to differential Galois theory, see [6]3 (taking into account the Fuchsianity of (I)). We first need a notation. For any $s \in \mathbb{P}^1_k$, denote by K_s the differential extension of $k[[x-s]]$ generated by $\log(x-s)$ and $(x-s)^k$. For any $f \in K_s$ we define the generalized order of f , and write $\text{g-ord}_s f$, as the minimum of the "exponents" α_i such that $f = \Sigma u_i(x-s)^{\alpha_i}\log(x-s)^{\nu_i}$ with $u_i \in k[[x-s]]\setminus 0$.

LEMMA. Let $\Lambda_r \in k(x)[d/dx]$ of order μ_r such that $\Lambda_r r = 0$. For any $s \in \mathbb{P}^1_k$, there exists μ_r solutions ${}^t(z_{s,i,0},\ldots,z_{s,i,\mu-1})$ of (I) in K_s^{μ} such that the μ_r elements $p(x,z_{s,i,0},\ldots,z_{s,i,\mu-1})$ of K_s span the kernel of Λ_r in K_s .

Sketch of proof of the theorem: let $\eta_{s,i}$ denote the exponents of Λ_r at s (i.e. the μ_r roots of its indicial polynomial). A simple computation of generalized orders based on the lemma gives:

$$(35) \quad \mathrm{Re}\sum_i \eta_{s,i} \begin{cases} \geqq \mu_r\varepsilon_s & \text{if } s\neq\infty \\ \geqq \mu_r(\varepsilon_\infty-N) & \text{if } s=\infty \\ \geqq \mathrm{ord}_0 r+(\mu_r-1)\varepsilon_0 & \text{if } s=0 \text{ because of the existence of the solution } Y \text{ of (I).} \end{cases}$$

An elementary computation of generalized orders of derivatives shows moreover that:

$$(36) \quad \text{g-ord}_s w \begin{cases} \geqq \mathrm{Re}\sum_i \eta_{s,i} - \frac{\mu_r(\mu_r-1)}{2} & \text{if } s\neq\infty \\ \geqq \mathrm{Re}\sum_i \eta_{s,i} + \frac{\mu_r(\mu_r-1)}{2} & \text{if } s=\infty , \end{cases}$$

where w denotes the Wronskian at the neighbourhood of s . Because of Liouville's equation:

(37) $d/dx\ w = (\mathrm{tr}\ \Gamma_k)w$, (where Γ_r denotes the matrix

$$\begin{pmatrix} 0 & 1 & & 0 \\ & & \ddots & \\ 0 & & & 1 \\ * & * & \cdots & * \end{pmatrix}$$

associated to Λ_r), we find

(38) $g\text{-}\mathrm{ord}_s w = \mathrm{residue}_s \mathrm{tr}\ \Gamma_r$.

On the other side, the residue formula over $\mathbb{P}^1_k$ reads:

(39) $$\sum_{\mathrm{Sing}\Gamma_r} \mathrm{residue}_s \mathrm{tr}\ \Gamma_r = 0 .$$

Putting (35),...,(39) together, one finds

$$\sum_{\substack{s \neq 0 \\ s \neq \infty}} \left(\mu_r \varepsilon_s - \frac{\mu_r(\mu_r-1)}{2}\right) + \mu_r(e_\infty - N) + \frac{\mu_r(\mu_r-1)}{2}$$

$$+ (\mu_r-1)\varepsilon_0 - \frac{\mu_r(\mu_r-1)}{2} + \mathrm{ord}_0 r \leq 0 \text{ , which}$$

gives (34) for $n = 1$, by remembering that only the elements of $\mathrm{Sing}\Gamma \setminus \{0,\infty\}$ can give a negative contribution to the first sum.

□

(1) D. Bertrand has informed me that J. Yebbou recently succeeded in making this method effective in the general case.

Chapter IV Fuchsian Differential Systems: Arithmetic Theory

§ 1. BACKGROUND OF P-ADIC ANALYSIS

We record here all the properties of p-adic analytic functions (and prove some of them) which will be used in the sequel of the book. We refer the reader to the general introduction [41], or to the more advanced booklet [26]. Everything becomes simpler if one restricts oneself, as we shall do, to the case of disks.

1.1 - Let k denote an algebraically closed field of characteristic 0, complete under a ultrametric (= non Archimedean = p-adic) absolute value $|\ |$, with residue field $\bar{k}$ of characteristic $p > 0$. This residue field is automatically algebraically closed, and the valuation group $\log_p|k|$ is dense in $\mathbb{R}$.
A typical example for k would be the field $\mathbb{C}_p$ (see general notations).
We remind that a series Σa_n converges in k if and only if its general term $a_n \longrightarrow 0$, and that $|\Sigma a_n| \leq \operatorname{Sup}_n |a_n|$. Let $r \in \mathbb{R}^+$, and $a \in k$; we denote by $D(a,r)$ the disk $\{x \in k;\ |x-a| < r\}$, which is an open and closed subset of k.
For $r = 1$, these disks are also called <u>residue classes</u>, for they correspond to the elements of $\bar{k} \cup \infty$ by reduction modulo the valuation ideal $D(0,1)$.

1.2 - <u>Gauss absolute value</u>
On $k[x]$, the Gauss absolute value is defined to be the maximum of the absolute value of the coefficients, and is still denoted by $|\ |$: $|\sum_{n=0}^{N} y_n x^n| := \operatorname{Max}_n |y_n|$.
Let $f,g \in k[x]$; Gauss' lemma shows that $|f|/|g|$ depends only

on $f/g \in k(x)$. This allows to extend $|\ |$ to an ultrametric absolute value on $k(x)$.

1.3 - Analytic elements

Let a be an element of k such that $|a| \leq 1$. According to M. Krasner one defines the ring E_a of analytic elements in $D(a,1)$ to be the completion (under the Gauss absolute value) of the ring of rational functions without pole in $D(a,1)$. This allows to consider analytic elements as functions on $D(a,1)$ in a natural way.

We also denote by E the completion of the whole ring $k(x)$ under $|\ |$; warning: this is much bigger than the fraction field $\overset{\circ}{E}_a$ of E_a . The residue field of all those ultrametric rings is $\bar{k}(x)$.

We record now two basic properties of analytic elements.

(1.3.1) E_a is a Noetherian principal Banach algebra over k ; any ideal is generated by a polynomial.

A consequence of (1.3.1) is that any analytic element in $D(a,1)$ vanishes only finitely many times when viewed as a function on this disk.

The second basic property is Krasner's principle of analytic extension:

(1.3.2) let $f \in E_0$, then for all but a finite number of residue classes $D(a,1)$, there exists a unique element of E_a which coincides with f as an element of E . At last we denote by E_{aa} the localization of E_0 at the ideal $(x-a)E_a$. By (1.3.1) we have $E_{aa} = E_a \otimes_{k[x]} k[x]_{(x-a)}$, where $k[x]_{(x-a)}$ denotes as in chapter 1 the localization of $k[a]$ at $(x-a)$.

REMARK. There is an isometric embedding of $(E,|\ |)$ in the (complete) Amice ring $\{f = \sum_{n\in\mathbb{Z}} f_n x^n$; $|f| := \mathrm{Sup}|f_n| < \infty$ and $\lim_{n\to-\infty} |f_n| = 0\}$; however we shall not use it in this book.

1.4 - Analytic functions

Let $a \in k$ and $r \in \mathbb{R}^+$. The ring of analytic functions in $D(a,r)$, denoted by $A(a,r)$, is the ring of convergent series $y = \sum_{n\geq 0} y_n(x-a)^n$ in $D(a,r)$ (under the Cauchy product); the requirement is that $\forall\rho < r$, $\lim_{n\to\infty}|y_n|\rho^n = 0$.

As in the classical case, one may expand y in Taylor series at any point of $D(a,r)$ and speek about zeroes and order of zeroes of y; $A(a,r)$ is stable under d/dx. For any $\rho < r$, one defines an absolute value $|\ |_a(\rho)$ on $A(a,r)$ by:

$$|\Sigma y_n(x-a)^n|_a(\rho) = \operatorname{Sup}_n |y_n|\rho^n .$$

It is easy to check, for any $f \in A(a,r)$, the formulae:

(1.4.1) $|f|_a(\rho) = \operatorname{Sup}_{x\in D(a,\rho)} |f(x)| = (\operatorname{Sup}_{|x|=\rho} |f(x)|$ if $\rho \in |k|)$.

For $\rho = r$, one sets furthermore $|f|_a(r) = \limsup_{p\to r} |f|_a \in [0,\infty]$, which is a multiplicative semi-norm.

For $|a| \leq 1$, we have $E_a \subset A(a,1)$ and $|f|_a(1) = |f|$ for any $f \in E_a$. Conversely, it is plain to check that any analytic function in $D(a,r)$ with $r > 1$ defines an element of E_a; warning: the ring of bound analytic functions on $D(a,1)$ under $|\ |_a(1)$ is much bigger than E_a.

1.5 - Meromorphic functions

The field $M(a,r)$ of meromorphic functions is the quotient field of $A(a,r)$. As in 1.2, one checks that $|\ |_a(\rho)$ extends to an ultrametric absolute value on $M(a,r)$ for $\rho < r$ (resp. to a multiplicative semi-norm for $\rho = r$), thanks to the rule $|f/g|_a(\rho) = |f|_a(\rho)/|g|_a(\rho)$. These (semi-) norms satisfy the following important inequality:

(1.5.1) $\forall f \in M(a,r)$, $\forall \rho \leq r$, $\forall n \in \mathbb{N}$, $\left|\frac{1}{n!}\frac{d^n}{dx^n} f\right|_a(\rho) \leq \rho^n |f|_a(\rho)$.

This is clear for any $f \in A(a,r)$ by looking at the Taylor expansion at a. The general case is settled by induction, using the Leibnitz formula:

$\frac{1}{n!}(f/g)^{(n)} = \frac{1}{n!} f^{(n)}/g - \sum_{m=0}^{n-1} \frac{1}{m!}(f/g)^{(m)} \frac{1}{(n-m)!} g^{(n-m)}/g$. Roughly speaking, a zero (resp. a pole) increases (resp. decreases) the norm; more precisely, for any $f \in M(a,r)$ and $r' < r'' \leq r$

(1.5.2) f has neither zero nor pole in the annulus $r' < |x| < r''$ if and only if $|f|_a(\rho)$ is constant in the range $\rho \in]r',r''[$.

At last, let us remark that $\partial = x\, d/dx$ is a contraction in $M(0,r)$ endowed with any of the $|\ |_0(\rho)$; this follows at once from (1.5.1).

Here is a diagram of inclusions between various rings that we considered:

$$\begin{array}{ccccccc} k[x] & \subset & E_0 & \subset & E_{00} & \subset & E \\ & & \cap & & \cap & & \\ & & A(0,r) & \subset & k[[x]] & & \text{for } 0 < r \leqq 1 . \\ & & \cap & & \cap & & \\ & & M(0,r) & \subset & k((x)) & & \end{array}$$

Here, $E, M(0,r)$ and $k((x))$ are fields; E_0 and E are complete rings under the Gauss absolute value.

§ 2. P-ADIC DIFFERENTIAL SYSTEMS

2.1 - Dwork-Frobenius lemma

LEMMA. *Let* $\Lambda = d^{\mu}/dx^{\mu} - \sum_{h=0}^{\mu-1} \gamma_h \, d^h/dx^h$ *be a differential operator with coefficients* $\gamma_h \in M(a,r)$ *. Assume that the kernel of* Λ *in* $M(a,r)$ *is of dimension* μ *. Then the* γ_h*'s are bounded in* $D(a,r)$ *; more precisely* $|\gamma_h|_a(r) \leqq r^{h-\mu}$ *for* $h = 0,\dots,\mu-1$.

Proof: let us denote by $y_0,\dots,y_{\mu-1}$ a basis of $\ker \Lambda$ in $M(a,r)$. We set $v_0 = y_0$

$$v_1 = d/dx(v_0^{-1}y_1)$$

$$\vdots$$

$$v_{\mu-1} = d/dx(v_{\mu-2}^{-1}d/dx(v_{\mu-3}^{-1}\dots d/dx(v_0^{-1}y_{\mu-1})\dots)) ,$$

and $\Lambda_j = d/dx + v_0\dots v_j \; d/dx\left(\frac{1}{v_0\dots v_j}\right) \in M(a,r)[d/dx]$, so that we have the so-called "Frobenius factorization"

$$\Lambda = \Lambda_{\mu-1}\circ\dots\circ\Lambda_1\circ\Lambda_0 .$$

Now inequality (1.5.1) gives the bound $1/r$ for $|\;|_a(r)$ applied to the constant term of Λ_j . The estimate $|\gamma_h|_a(r) \leqq 1/r^{\mu-h}$ follows at once from the factorization, and inequality (1.5.1) again.

□

COROLLARY. *Let* $\partial^{\mu} y = \sum_{h=0}^{\mu-1} g_h \partial^h y$ *be a differential equation with coefficients in* $M(0,r)$ *and which admits a fundamental basis of solutions in* $M(0,r)$. *Then* $|g_h|_0(r) \leq 1$, *for* $h = 0,\dots,\mu-1$.

2.2 - *Norms on matrix rings*

It is convenient to extend the definition of the absolute values $|\ |$, respectively $|\ |_a(\rho)$, to the case of matrices. We get norms on spaces of matrices by setting:

$$\|H\| = \operatorname{Max}_{i,j} |H_{ij}| \quad \text{for} \quad H \in M_\mu(E)$$

$$\|H\|_a(\rho) = \operatorname{Max}_{i,j} |H_{i,j}|_a(\rho) \quad \text{for} \quad H \in M_\mu(M(a,r)) \quad \text{and} \quad \rho < r .$$

Then $M_\mu(E_a)$ and $M_\mu(E)$ become Banach k-algebras under $\|\ \|$ (for any $a \in k$ such that $|a| \leq 1$).

Similarly one defines a semi-norm on $M_\mu(M(a,r))$ by

$$\|H\|_a(r) = \operatorname{Max}_{i,j} |H_{i,j}|_a(r) .$$

REMARK. Let $G \in M_\mu(E)$ and $H \in GL_\mu(E)$. Since d/dx and ∂ are contractions of $(k(x),|\ |)$ they extend to derivations on E . Thus ∂H and $H[G]$ are well-defined, and

$$\|H[G]\| \leq \operatorname{Max}(\|H\|\ \|H^{-1}\|\ \|G\| , \|\partial H\|\ \|H^{-1}\|) \leq$$

$$\leq \|H\|\ \|H^{-1}\| \operatorname{Max}(1,\|G\|) .$$

2.3 - *Estimates for* $G_{[n]}$

Let G be a matrix in $M_\mu(M(a,r))$, *with* $|a| \leq 1$. We use the notation $G_{[n]}$ of III (8).

PROPOSITION. *Assume that the entries of* G *are bounded, and that the system* $\partial X = GX$ *has a solution* X_a *in* $GL_\mu(M(a,r))$. *Then for every* $n \in \mathbb{N}$, $\|G_{[n]}\|_a(r) \leq \operatorname{Max}(1,(\|G\|_a(r))^{\mu-1})$.

Proof: we shall use a method of E. Bombieri in order to deduce this statement from the previous lemma. Since k is algebraically closed, we may find a sequence $\alpha_n \in k$ with $|\alpha_n| \to r^-$ and then the change of variable $x \longmapsto \alpha_n(x-a) + a$ enables us to consider only the case $r = 1$, $a = 0$.

The construction III 3.2 then applies: we get some elements ${}_{nj}S$ in $M(0,r)$, and a matrix $S = H^{-1} \in GL_\mu(M(0,r))$, satisfying the equations

$$x^{-n}(H[G])_{[n]}\, SX_a = d^n/dx^n(SX_a) = ({}_{ij}S)_{\substack{i=n,\dots,n+\mu-1\\ j=0,\dots,\mu-1}} \cdot X_a ,$$

whence ${}_{nj}S = x^{-n}{}_{0j}((H[G])_{[n]}S)$. It follows that

$|{}_{nj}S|_0(1) \leq (\mathrm{Max}|\alpha_i|)\cdot \mathrm{Max}_{m<\mu} \| G_{[m]}\|_0(1)$. Remind the definition of ${}_{nj}S$:

$${}_{nj}S = \sum_{m=0}^{n} \sum_{i=0}^{\mu-1} \binom{n}{m} \alpha_i x^{m-n} \frac{d^m x^{l_i}}{dx^m} \cdot {}_{ij}G_{[n-m]} ,$$ and choose for any fixed i_0 : $\alpha_{i_0} = 1$, $\alpha_i \to 0$ for $i \neq i_0$ and $l_{i_0} = 0$. This gives

$$\left|{}_{i_0 j}G_{[n-m]}\right|_0(1) \leq \mathrm{Max}_{m<\mu} \|G_{[m]}\|_0(1)$$

$$\leq \mathrm{Max}(1, \|G\|_0(1))$$ by induction on formula (8).

□

2.4 - <u>Generic disks</u>

Since the early works of B. Dwork on p-adic differential equations, one knows that a convenient way to encode information about such an equation is, roughly speaking, by transfering it into a "generic disk", which has nice permanence properties; transfer principles then help to bring it back into special disks. We shall study such a process in the next chapter.

A generic point ,often denoted by t , is an element of an algebraically closed and complete extension Ω of the valued field k linearly disjoint from $k((x))$, such that $|t-a| = 1$ for any $a \in k$ with $|a| \leq 1$. The image $\bar{t}$ of t in the residue field of Ω is then transcendental over $\bar{k}$. Note that t^n is another generic point, for any non-zero integer n . The generic disks $D(t,R)$ are those of Ω , and we extend the previous notations by replacing k by Ω ; whether we use that extension or not should be clear according to the context. For instance, the field E can be embedded into the ring of analytic elements in $D(t,1)$; this provides an isometric embedding $(E,|\ |) \hookrightarrow (A(t,r),|\ |_t(r))$ for any $r < 1$, and $|f| = |f(t)|$ for any $f \in E$. This allows us also to apply proposition 2.3 for $a = t$ when $G \in M_\mu(E)$.

2.5 - <u>Generic radius of solvability</u>

The definitions of chapter III are used up to now.

DEFINITION. Let M be a ∂-module over E ; its generic radius of solvability, denoted by $R(M)$, is the greatest real number $R \leq 1$ such that M is solvable in $A(t,R)$.

By definition of the matrices $G_{[n]}$ and by Taylor's formula, we get the following expansion for a solution X_t in $M_\mu(A(t,R))$ of the system $\partial X = GX$ associated with some basis of M :

$$X_t = \sum_{n\geq 0} \frac{(x-t)^n}{n!t^n} G_{[n]}(t)X_t(t) .$$

Since $\| G_{[n]}(t) \| = \| G_{[n]} \|$, Hadamard's formula yields:

$$(2.5.1) \quad R(M) = \operatorname{Min}\left(\varliminf_{n\to\infty} \left\| \frac{G_{[n]}}{n!} \right\|^{-1/n} , 1 \right) .$$

LEMMA. $R(M) > 0$; more precisely, if $G \in M_\mu(E)$ represents M in some basis, then $R(M) \geq |p|^{\frac{1}{p-1}} \cdot \operatorname{Min}(1, \| G\|^{-1})$.

Proof: by the recursion (8), we get $\| G_{[n]} \| \leq \| G\|^n$. On the other side $|n!|^{-1/n} \leq |p|^{\frac{-1}{p-1}}$, according to I, appendix. Hence the required inequality follows from (2.5.1).

□

PROPOSITION 1. Let M^* be the dual ∂-module of M ; then $R(M^*) = R(M)$. If there is an exact sequence $0 \to M' \to M \to M'' \to 0$, then $R(M) = \operatorname{Min}(R(M'),R(M''))$.

Proof: let G and X_t as above. By III 1.3, one has to prove that $X_t^{-1} \in M_\mu(A(t,R))$. Let $w = \det X_t$ be the Wronskian determinant, which satisfies $\partial w = \operatorname{tr} G w$; since $\operatorname{tr} G \in A(t,R)$, w is invertible in $A(t,R)$, and this is enough to conclude that $R(M^*) \geq R(M)$. The equality follows by symmetry, since $M^{**} \simeq M$.
It is obvious that any quotient of a ∂-module solvable in $A(t,R)$ is solvable in $A(t,R)$; this remains true for a submodule by duality using $R(M^*) = R(M)$. This proves the inequality: $\operatorname{Min}(R(M'),R(M'')) \geq R(M)$.
Conversely let us write $R = \operatorname{Min}(R(M'),R(M''))$. Then we have isomorphisms

$$\left.\begin{array}{l} M' \otimes_E \ A(t,R) \simeq A(t,R)^{\mu'} \\ M'' \otimes_E \ A(t,R) \simeq A(t,R)^{\mu''} \end{array}\right\} \text{ with trivial connection.}$$

Thus in some suitable basis, the matrix representing M looks like $G = \begin{pmatrix} 0 & G_1 \\ 0 & 0 \end{pmatrix}\begin{matrix}\}\mu'' \\ \}\mu'\end{matrix}$. A solution X_t of $\partial X = GX$ at t is given by $\begin{pmatrix} I & \int \frac{G_1}{x} \\ 0 & I \end{pmatrix}$. Hence $X_t \in GL_\mu(A(t,R))$, and it follows that $R(M) \geq R$. □

(2.5.2) CONVENTION. Assume here that $|\ |$ is an Archimedean absolute value, in contrast to our previous hypotheses. Let M be a ∂-module over $k(x)$. <u>We then set</u> $R(M) = 1$.

Explanation: by Gelfand-Mazur theorem, k is isometric to $\mathbb{C}$ in our case; by analogy with the previous case, where $D(t,1)$ was an ordinary disk, Cauchy's theorem would imply that $R(M) = 1$.

Using the idea of Bombieri-Sperber, we give a local version of a result of N. Katz which relates the generic radii of solvability at all the finite places (in the global case) to the singularity structure and to the exponents, see [39] and [8]. Let $\Lambda \in E_0[1/x, d/dx]$ be a differential operator of order μ with coefficients in $E_0[1/x]$, say $\Lambda = \frac{1}{\mu!}\frac{d^\mu}{dx^\mu} - \sum_{j=0}^{\mu-1}\gamma_{\mu j}\frac{1}{j!}\frac{d^j}{dx^j}$. We define λ_j, ρ, J and $\tilde{\Lambda}$ as in III 4; e.g.
$\gamma_{\mu j} = \frac{\lambda_i}{x^{\delta_j}} + \text{h.o.t.}$, etc.... . We also write $R(\Lambda)$ for the generic radius of solvability of the ∂-module associated to Λ as in chapter III.

PROPOSITION 2. <u>One has</u> $R(\Lambda) \leq |p|^{1/p-1}$ <u>in each of the following cases</u>:

i) $\rho > 1$ (irregular case) <u>and</u> $|\lambda_j| \geq 1$ <u>for all</u> $j \in J$

ii) $\rho = 1$ (regular case) <u>and at least one of the exponents of</u> Λ (<u>at</u> 0) <u>is at distance</u> ≥ 1 <u>from</u> $\mathbf{Z}_p$.

Proof: let $R(0,\tilde{\Lambda})$ denote the greatest real number $r \leq 1$ such that $\tilde{\Lambda}$ is solvable in $A(0,r)$. By Lemma III 4 and Hadamard's formula, we have $R(0,\tilde{\Lambda}) = \underset{j}{\mathrm{Min}}(1, \varliminf_{m\to\infty} |\lambda_{m,j}|^{-1/m})$ where $\lambda_{m,j}$

are defined by $\gamma_{m,j} = \dfrac{\lambda_{m,j}}{x^{\rho(m-j)}} + \text{h.o.t.}$ as in loc. cit.; note that $\gamma_{m,j} \in E_0[1/x]$. By factoring out the possible pole 0, we find $|\gamma_{m,j}| = |x^{\rho(m-j)}\gamma_{m,j}| = \overline{\lim_{r\to 1}} \sup_{n \geq \rho(j-m)} |\gamma_{m,j,n}| r^{n+\rho(m-j)}$

$$\geq |\lambda_{m,j}| .$$

It follows that $R(\Lambda) \leq R(0,\tilde{\Lambda})$. Hence it suffices to show that $R(0,\tilde{\Lambda}) \leq |p|^{\frac{1}{p-1}}$ in both cases i) and ii), which we shall treat separately.

Case i): $\tilde{\Lambda}$ is a constant coefficient differential operator and a basis of $\ker \tilde{\Lambda}$ is $x^{\nu}\exp \alpha x$ where α runs over the roots of the characteristic polynomial $\frac{x^{\mu}}{\mu!} - \sum_{j\in J} \lambda_j \frac{x^j}{j!}$ and $\nu = 0,1,\ldots,\nu_{\alpha}$ with ν_{α} = multiplicity of α. Because $\rho > 1$, J is non-empty and all $|\lambda_j|$ are bound from below by 1 for $j \in J$. This implies that the absolute values of the roots α are also bound from below by 1. Since the radius of convergence of $\exp \alpha x$ is $|\alpha|^{-1}|p|^{1/p-1}$, we get $R(0,\tilde{\Lambda}) \leq |p|^{\frac{1}{p-1}}$ as required.

Case ii): now a basis of $\ker \tilde{\Lambda}$ is $(1+x)^{\alpha}(\log(1+x))^{\nu}$ where α runs over the exponents of $\tilde{\Lambda}$ at $x = -1$ (which are the exponents of Λ at 0) and $\nu = 0,\ldots,\nu_{\alpha}$ with ν_{α} = multiplicity of α.
Since at least one α is at distance ≥ 1 from $\mathbf{Z}_p$, it follows that the nth coefficient $\frac{(-1)^n}{n!}(-\alpha)_n$ of $(1+x)^{\alpha}$ has absolute value $|n!|^{-1}\text{Max}(1,|\alpha|^n)$; hence the radius of convergence of $(1+x)^{\alpha}(\log(1+x))^{\nu}$ is again $\leq |p|^{\frac{1}{p-1}}$, and it is plain to deduce that $R(0,\tilde{\Lambda}) \leq |p|^{\frac{1}{p-1}}$.

§ 3. GLOBAL RADIUS OF A ∂-MODULE OVER $K(x)$

3.1 - We now turn back to the arithmetic setting of chapter I. In particular, K denotes a number field, etc. ... To each finite place $v \in \Sigma_f$, we associate the ultrametric field $k_v = \mathbb{C}_v$ which satisfies our hypotheses; we have $k_v \supset K_v$. A

∂-module $\mathcal{M}$ over $k(x)$ gives rise to a collection $\{\mathcal{M}_v = \mathcal{M} \otimes_K k_v\}$ of ∂-modules over $k_v(x)$. In order to point out the dependence on v, we shall add a subscript v to all objects defined in §§ 1-3 from $k_v, \mathcal{M}_v$. Since $k_v(x) \subset E_v$, the generic radii of solvability are defined. Sometimes, we shorten the notation $R(\mathcal{M}_v)$ into R_v, for some fixed $\mathcal{M}$ as above. When no confusion results, we shall often write simply p instead of $p(v)$ for the residue characteristic.

3.2 - REMARK. For any polynomial $g \in K[x] \setminus \{0\}$, we have $|g|_v = 1$ for almost all $v \in \Sigma_f$; this carries over for any $g \in K(x) \setminus \{0\}$ and extends to matrices: every $H \in GL_\mu(K(x))$ satisfies $\| H \|_v = \| H^{-1} \|_v = 1$ for almost $v \in \Sigma_f$.

3.3 - With the convention (2.5.2), we may write

$$\sum_{v \in \Sigma(K)} \log^+ 1/R_v \quad \text{for} \quad \sum_{v \in \Sigma_f} \log 1/R_v .$$

In analogy with chapter I, and following Bombieri [7], we define the global radius of a ∂-module $\mathcal{M}$ over $K(x)$ by the formula

$$\rho(\mathcal{M}) := \sum_{v \in \Sigma(K)} \log^+ 1/R_v(\mathcal{M}) \in [0,\infty] .$$

Let $\{m_i\}_{i=0}^{\mu-1}$ be a $K(x)$-basis of $\mathcal{M}$, and let G be the matrix which represents ∂ with respect to this basis (in the sense of 2.1). Pushing further the analogy with chapter I, we define a collection of sequences of non-negative real numbers:

$$h_{v,n}(\mathcal{M},\{m_i\}) = \begin{cases} 0 & \text{if } v \in \Sigma_\infty \\ \frac{1}{n} \log^+ \operatorname{Max}_{m \leq n} \left\| \frac{G_{[n]}}{n!} \right\|_v & \text{if } v \in \Sigma_f . \end{cases}$$

Let $X_{t_v} \in GL_\mu(\mathcal{A}(t,R_v))$ a solution of the system $\partial X = GX$ with $X_{t_v}(t_v) = I$. By looking at the coefficients of X_{t_v}, one may write $h_{v,n}(\mathcal{M},\{m_i\})$ in the following manner, for $v \in \Sigma_f$:

(3.3.1) $\quad h_{v,n}(\mathcal{M},\{m_i\}) = \frac{1}{n} \log^+ \operatorname{Max}_{\substack{m \leq n \\ i \leq \mu \\ j \leq \mu}} \left| {}_{ij}X_{t_v,n} \right|_v$, where the subscript n refers to the nth-coefficient at t_v.

LEMMA 1. <u>One has</u> $\rho(M) = \sum_v \overline{\lim_{n\to\infty}} h_{v,n}(M,\{m_i\})$; $\rho(M)$ <u>is invariant under finite extension of</u> K.

Proof: Taking into account (2.5.1), the argument given in the proof of Lemma 1 of I carries over, and yields $\log 1/R_v = \overline{\lim_{n\to\infty}} h_{v,n}(M,\{m_i\})$. We leave the second assertion to the reader. □

LEMMA 2. <u>One has the following rules:</u>

a) $\rho(M) = \rho(M^*)$

b) <u>If there is an exact sequence</u> $0 \to M' \to M \to M'' \to 0$, <u>then</u>
$$\mathrm{Max}(\rho(M'),\rho(M'')) \leq \rho(M) \leq \rho(M' \oplus M'') \leq \rho(M') + \rho(M'')$$

c) <u>for any positive integer</u> N, $\rho(M^{\odot N}) \leq \rho(M)$

d) $\rho(\det M) \leq \rho(M)$.

Proof: a) and b) follow from proposition 1 of 2.5; c) follows from the fact that a basis of solutions of $M^{\odot N}$ into $A(t,R)$ is obtained from a basis of solutions of M into $A(t,R)$ by taking products. A similar argument involving the determinant yields d). □

For a differential system or equation L attached to a basis $\{m_i\}$ (resp. a cyclic basis) of M , we shall sometimes write $\rho(L)$ instead of $\rho(M)$; also, $h_{v,n}(L)$ for $h_{v,n}(M,\{m_i\})$ ...

§ 4. <u>SIZE OF A ∂-MODULE</u>

4.1 - We carry on developing the analogy with chapter 1; in the next two chapters, this will become much more than a formal analogy (see also 5.4 below).
We define the <u>size</u> of a ∂-module M over $K(x)$ by the formula:

$$\sigma(M) := \overline{\lim_{n\to\infty}} \sum_{v\in\Sigma(K)} h_{v,n}(M,\{m_i\}) .$$

LEMMA 1. <u>This definition does not depend on the choice of the basis</u> $\{m_i\}$, <u>and is invariant under finite extension of</u> K .

Proof: the second assertion is routine; let us prove the first one. According to III 2.4 a change of basis is described by some $H \in GL_\mu(K(x))$: $X_{t_v} = HX'_{t_v}$. We have remarked in 3.2 that

$\|H\|_v = \|H^{-1}\|_v = 1$ for all but a finite number of places v, and $\sup_n \|(H(x+t_v))_n\| < \infty$ for all v (since rational functions are bounded analytic functions in the generic disk). Making use of formula (3.3.1), it follows that
$h_{v,n}(M,\{m_i'\}) \leq h_{v,n}(M,\{m_i'\}) + \frac{c_v}{n}$ for some constants c_v which are 0 for almost every v. This implies the independence of σ from $\{m_i\}$. □

LEMMA 2. <u>The following inequalities hold:</u>

a) $\sigma(M' \oplus M'') \leq \sigma(M') + \sigma(M'')$

b) <u>for any subquotient</u> M' <u>of</u> M, $\sigma(M') \leq \sigma(M)$

c) <u>for any positive integer</u> N, $\sigma(M^{\odot N}) \leq \sigma(M)(1+\log N)$.

Proof: a) is straightforward. For b) it suffices to note that if $0 \to M' \to M \to M'' \to 0$ is an exact sequence, a solution at t_v of the associated differential system in a suitable basis can be written in the shape

$$X_{t_v} = \begin{pmatrix} X''_{t_v} & X^0_{t_v} \\ 0 & X'_{t_v} \end{pmatrix}$$

, with X''_{t_v} resp. X'_{t_v} corresponding to solutions of M'' resp. M'.

Similarly a solution corresponding to $M^{\odot N}$ at t_v can be written $X^{\odot N}_{t_v}$, (symmetric powers); c) follows from an adaption of Shidlovski's argument given in the proof of I.1. Lemma 2; details are left to the reader. □

4.2 - The example of polylogarithmic differential equations I 5.3 shows that one cannot expect an equality like 3.3 b) for the size in the case of an arbitrary extension. Nevertheless the behaviour of σ can be controled.

PROPOSITION. <u>If there is an exact sequence</u> $0 \to M' \to M \to M'' \to 0$, <u>then</u> $\sigma(M) \leq 1 + 2\sigma(M' \oplus M'' \oplus M''^*)$.

Proof: as said before, we may choose a suitable basis so that the representative matrix G, and a solution X_{t_v} at t_v of $\partial X = GX$, look like:

$G = \begin{pmatrix} G'' & G^0 \\ 0 & G' \end{pmatrix}$, $X_{t_v} = \begin{pmatrix} X''_{t_v} & X^0_{t_v} \\ 0 & X'_{t_v} \end{pmatrix}$. From the formula

$\partial X^0_{t_v} = G'' X^0_{t_v} + G^0 X'_{t_v}$, we deduce that

$X^0_{t_v} = X''_{t_v}\left(\int_t^x X''^{-1}_{t_v}\left(\frac{G^0}{x}\right)X'_{t_v} + A\right)$, with $A \in M_\mu(\Omega_v)$. Since we want that $X_{t_v}(t_v) = I$, we have to put $A = I$. For notational convenience, we now omit the subscript t_v . We get

$$_{ij}X_n = \sum_{l_1,l_2,l_3} \sum_{\substack{\sum_{i=0}^{3} m_i = n+1 \\ m_0 \leq n}} \frac{1}{m_1+m_2+m_3}\ {}_{il_1}X''_{m_0}\ {}_{l_1l_2}(X''^{-1})_{m_1}\ {}_{l_2l_3}\left(\frac{G^0}{x}\right)_{m_2}\ {}_{l_3j}X'_{m_3} .$$

But since ${}_{l_2l_3}\left(\frac{G^0}{x}\right) \in K(x) \subset E_v$, we have

$\left|{}_{l_1l_2}\left(\frac{G^0}{x}\right)_{m_2}\right|_v \leq \left\|\frac{G^0}{x}\right\|_{t_v}(1) = \|G^0\|_v$, so that

$$|{}_{ij}X_n|_v \leq |\mathrm{l.c.m}(1,2,\dots n+1)|_v^{-1}\,\|G^0\|_v \max_{\substack{l_1,l_2,l_2 \\ m_{l_1}+m_{l_2}+m_{l_3} \leq n+1}} |{}_{il_1}X''_{m_{l_1}}\ {}_{l_1l_2}(X''^{-1})_{m_{l_1,l_2}}\ {}_{l_3j}X'_{m_{l_3}}|_v .$$

By reordering ${}_{il_1}X''$, ${}_{l_1l_2}(X''^{-1})$ and ${}_{l_3j}X'$ we may suppose that $m_{l_1} \geq m_{l_1,l_2} \geq m_{l_3}$, hence $m_{l_1} + m_{l_1,l_2} + m_{l_3} \leq (1+1/2+1/3)(n+1) < 2(n+1)$. We find, for the chosen bases on M, M', M'' and the dual one on M''^* :

$$h_{v,n}(M) < 2\left(\frac{n+1}{n}\right)h_{v,n+1}(M'\oplus M''\oplus M''^*) + \frac{1}{n}\log\|G^0\|_v + \frac{1}{n}\log|\mathrm{g.c.m}(1,\dots,n+1)|_v^{-1} .$$

Because $\|G^0\|_v = 1$ for almost every v , and $\|G\|_v < \infty$ for all v,

$$\sum_v h_{v,n}(M) < \left(2+\frac{2}{n}\right)\sum_v h_{v,n+1}(M'\oplus M''\oplus M''^*) + \frac{1}{n}(\log \mathrm{g.c.m}(1,\dots,n+1) + \mathrm{constant})$$

$$\leq 2\sum_v h_{v,n+1}(M'\oplus M''\oplus M''^*) + 1 + o(n)$$ by the prime number theorem,

and the required inequality follows.

□

For a differential system or equation L, we allow the notation $\sigma(L)$ for the size of the associated ∂-module over $K(x)$.

OPEN QUESTION. It is easy to see that a trivial ∂-module M over $K(x)$ has vanishing size; does the converse statement hold?

§ 5. G-OPERATORS

5.1 - Bombieri's and Galochkin's condition

Let M be a ∂-module of rank μ over $K(x)$. The condition $\rho(M) < \infty$ can be restated as follows:

Bombieri's condition: $\prod R_v > 0$, see [7] 10.

It means that the generic radii of solvability R_v cannot be too small. Let us now consider the condition $\sigma(M) < \infty$; it corresponds, for the case of ∂-modules, to the condition which defines G-functions. Let us choose some basis, so that M is represented by $G \in M_\mu(K(x))$. Let $U \in \mathcal{O}_K[x]$ denote a common denominator for the entries ${}_{ij}G$ of G. Then $\sigma(M) < \infty$ can be written: there exists a constant C so that

$$\prod_{v\in\Sigma_f} \operatorname*{Max}_{m\leq n} \left\| \frac{U^m G_{[m]}}{m!} \right\| \leq C^m .$$

This is clearly equivalent to:

Galochkin's condition: the common denominator $d_n \in \mathbb{N}$ of the coefficients of $\frac{U^m G_{[m]}}{m!}$ for $m = 0,\dots,n$ is bounded above by a geometric progression in n, see [30] [60].

5.2 - THEOREM. The following inequalities hold $\rho(M) \leq \sigma(M) \leq \rho(M) + \mu-1$; they imply the equivalence of Bombieri's and Galochkin's conditions[1].

Proof: the second inequality is implicit in [7] 6 and uses the theorem of Dwork-Robba (see Appendix). We pick a cosingularity a of M in K, then we make the change of variable $x \longmapsto x-a$ which enables us to reduce to a differential equation according to III 3.2, and by $x \longmapsto x+a$ we go back to the original

---

(1) E. Bombieri informed me that there is recent independent work of Chudnovsky in this direction.

variable; hence we gave got a differential operator in $K(x)[d/dx]$ which represents $\mathcal{M}$, and the corresponding matrix G looks like:

$$\begin{pmatrix} 0 & x & & \bigcirc \\ & & \ddots & \\ \bigcirc & & \ddots & x \\ ** & \cdots & \cdots & * \end{pmatrix}$$

. Put $\{n,\mu-1\}_v := \operatorname{Max}\{|l_1 \cdots l_{\mu-1}|_v^{-1} ;$ $0 < l_1 < \ldots < l_{\mu-1} \leq n;\ l_1,\ldots,l_{\mu-1} \in \mathbb{N}\}$. After Dwork-Robba, we then have:

$$\left\|\frac{G_{[n]}}{n!}\right\|_v = \operatorname*{Max}_{i,j} \left|\frac{{}_{ij}G_{[n]}}{n!}\right|_{t_v} (R_v) \leq \{n,\mu-1\}_v R_v^{-n} .$$

Hence $h_{v,n} \leq \frac{1}{n}\log\{n,\mu-1\}_v + \log 1/R_v$. By summing over $v \in \Sigma_f$ and taking $\overline{\lim}$ over n , one gets

$$\sigma(\mathcal{M}) \leq \rho(\mathcal{M}) + \overline{\lim_{n\to\infty}} \frac{1}{n} \sum_{v\in\Sigma_f} \log\{n,\mu-1\}_v .$$

We may simplify this inequality by noting that, thanks to our normalizations, $\{n,\mu-1\}_v \leq p^{(d_v/d)\cdot(\mu-1)[\log n/\log p]}$, whence

$$\sum_{v|p} \log\{n,\mu-1\}_v \leq (\mu-1)\log p \sum_{v|p} \frac{d_v}{d} [\log n/\log p]$$

$$= (\mu-1)(\log p) v_p(\text{l.c.m.}(1,2,\ldots,n))$$

which implies $\sum_{v\in\Sigma_f} \log\{n,\mu-1\}_v \leq (\mu-1)\log \text{l.c.m}(1,2,\ldots,n)$, and $\overline{\lim_{n\to\infty}} \frac{1}{n} \sum_{v\in\Sigma_f} \log\{n,\mu-1\}_v \leq \mu-1$ follows from the prime number theorem. For the other inequality, let us recall that $\rho(\mathcal{M}) = \sum_v \overline{\lim_v}\, h_{v,n}$ and $\sigma(\mathcal{M}) = \overline{\lim_n} \sum_n h_{v,n}$, with $h_{v,n} \geq 0$. By Fatou's lemma , it is enough to prove that $\lim_n h_{v,n}$ exists for every v : indeed $\sum_v \lim_n h_{v,n} \leq \lim_n \sum_v h_{v,n} \leq \sigma(\mathcal{M})$. Let us put $\Gamma_m = \frac{x^{-m}G_{[m]}}{m!}$ and $g_{v,m} = \frac{1}{m}\log^+ \|\Gamma_m\|_v$, so that $h_{v,n} = \operatorname*{Max}_{m\leq n} \frac{m}{n} g_{v,m}$. Note that Γ_m "represents" $\frac{1}{m!} d^m/dx^m$ on a solution X of $\partial X = GX$; let us look at $\frac{1}{m!n!} \frac{d^{m+n}}{dx^{m+n}} X = \frac{1}{m!} d^m/dx^m(\Gamma_n X)$. By Leibnitz's rule, the following identity holds true:

$$\binom{m+n}{n} \Gamma_{m+n} = \sum_{h=0}^{m} \frac{1}{h!}\left(\frac{d^h}{dx^h} \Gamma_n\right)\Gamma_{m-h} \text{ , which yields}$$

$$(*) \quad (m+n)g_{v,m+n} \leq \operatorname*{Max}_{h\leq m}(h\, g_{v,h}) + n\, g_{v,n} - \log\left|\binom{m+n}{n}\right|_v .$$

By induction, we get

$$g_{v,hm} \leq \underset{1\leq m}{\operatorname{Max}} \frac{1}{m} g_{v,1} - \sum_{l=1}^{h-1} \frac{1}{hm} \log\left|\binom{lm+m}{lm}\right|_v .$$

But we have $-\sum_{l=1}^{h-1} \frac{1}{hm} \log\left|\binom{lm+m}{lm}\right|_v = \frac{1}{m}(h^{-1}\log|h\ m!|_v^{-1}$

$+ \log|m!|_v) \leq \frac{d_v}{d} \frac{\log p}{(p-1)p^N}$ if $m \geq p^N$ for some

$N \in \mathbb{N}$. Another application of (*) yields

$$g_{v,hm+n} \leq \underset{1\leq n}{\operatorname{Max}} \frac{1}{hm} g_{v,1} + g_{v,hm} - \frac{1}{hm+n}\log\left|\binom{hm+n}{n}\right|_v \quad \text{for} \quad 0\leq n<m ,$$

$$\leq (1+h^{-1})\left[\underset{1\leq m}{\operatorname{Max}}\left(\frac{1}{m}g_{v,1}\right)\right] + \frac{d_v}{d}\frac{\log p}{p^N(p-1)} - \frac{1}{hm+n}\log\left|\binom{hm+n}{n}\right|_v ,$$

$$\leq (1+h^{-1})\left[\underset{1\leq m}{\operatorname{Max}}\left(\frac{1}{m}g_{v,1}\right) + \frac{d_v}{d}\frac{\log p}{p^N(p-1)}\right] ,$$

whence $h_{v,n} \leq \left(1+\left[\frac{n}{m}\right]^{-1}\right)\left(h_{v,m} + \frac{d_v}{d}\frac{\log p}{p^N(p-1)}\right)$, when $n\geq m\geq p^N$.

It follows that the bounded set $\{h_{v,n}\}_{n\geq 0}$ has only one limit point in $\mathbb{R}$, hence the sequence $h_{v,n}$ converges. The proof of the theorem is now complete. □

Since these conditions appeared earlier independently in the literature, we are faced to the problem of choosing a terminology. I am not convinced by Bombieri's proposal to call such differential operators "of arithmetic type" [7] 10, and shall rather follow Debes' terminology "G-operator" [20].

Hence we shall call <u>G-operator</u> any differential operator $\partial - G$ or $d^\mu/dx^\mu - \sum \gamma_j \frac{d^\mu}{\partial x_j}$ with <u>finite</u> global radius (or equivalently finite size), with the following motivations: firstly, it stresses the link with G-functions (see next chapters); secondly, it reminds Galochkin's condition, and the typical Geometric growth of denominators; thirdly, it will remind a tantalizing conjecture of Dwork-Bombieri according to which G-operators should come from Geometry (in the sense of chap. II ; see below V, Appendix for a discussion).

REMARK about the proof of 5.2. In § 6 of next chapter, we give a new (independent) proof of the implication $\rho(M) < \infty \Rightarrow \sigma(M) < \infty$, which does not use Dwork-Robba' theorem. Anyway, the precise estimate $\sigma(M) \leq \rho(M) + \mu - 1$, which is best possible in general (look at polylogarithms!), will not be needed.

5.3 - Katz's theorem. The result of Katz which was mentioned in 2.5 may be stated as follows:

THEOREM (Katz). Let M be a ∂-module over $K(x)$ such that $\rho(M) < \infty$. Then M is Fuchsian with only rational exponents at every singularity in $\mathbb{P}^1_{\bar{\mathbb{Q}}}$. The term "Fuchsian" deserves an explanation: one claims the existence of a $\bar{\mathbb{Q}}[[x-s]]$-lattice in $M \otimes \bar{\mathbb{Q}}((x-s))$ stable under $(x-s)d/_{d(x-s)}$ for any $s \in \mathbb{P}^1_{\bar{\mathbb{Q}}}$; see also III appendix for an equivalent statement (the equivalence is classical, see e.g. [21]).

Proof: let us first remark that $\prod R_v > 0$ implies that $R_v > |p|^{1/p-1}$ if $v|p$ for any p in some set P_1 of prime numbers which has Dirichlet density 1. Indeed, denoting by P_0 the complement of P_1 , we have

$$\prod_{v|p\in P_0} R_v > 0 \Rightarrow \sum_{p\in P_0} \frac{\log p}{p-1} < \infty \Rightarrow \sum_{p\in P_0} \frac{1}{p} < \infty \Rightarrow \text{density } P_0 = 0 .$$

We reduce to a differential equation according to 3.2, cf. the discussion in the proof of the previous theorem; hence we have got a differential operator Λ in $K(x)[d/dx]$ which represents M . We add to P_0 the finite number of primes p such that $|\lambda_j|_v < 1$ for some $v|p$ and some $j \in J$, where the λ_j's $\in K \setminus \{0\}$ have the same meaning as in III 4 or IV 3. Let us remark that the statement we want to prove is invariant under finite extension of K , hence we may and shall assume that $\text{Sing}\,\Lambda \subseteq \mathbb{P}^1_K$. Once again, one adds to P_0 the finite number of primes p such that $|s_1|_v > 1$ or $|s_1-s_2|_v < 1$ for some $v|p$ and some $s_1 \neq s_2 \in \text{Sing}\,\Lambda \setminus \{0\}$. Then for any $v|p \in P_1$, any $s \in \text{Sing}\,\Lambda$, and any generic point t_v in some extension Ω_v of k_v, $t_v + s$ remains a generic point (as usual, we set $t_v \pm \infty = 1/t_v$). Hence the change of variable $x \longmapsto x-s$ does not modify R_v , and it suffices to prove the statement for $s=0$. Now by construction of P_1 we have for any $v|p \in P_1$:

$$\Lambda \in E_{0v}[1/x, d/dx],\ |\lambda_j|_v \geq 1 \text{ for } j \in J,\ R_v > |p|_v^{\frac{1}{p-1}} .$$

Proposition 2 of 2.5 applies: case i) cannot occur, so that 0 is a logarithmic singularity ($\text{ord}_0\, \gamma_{\mu,j} \geq j-\mu$) ; case ii) cannot occur, so that the reduction mod.v of the exponents at 0 belong to $\mathbb{F}_p$. Since density $P_1 = 1 > 1/2$, Chebotarev's theorem tells us that the exponents must belong to $\mathbb{Q}$.

□

REMARK. This proof shows that we could replace the condition $\prod R_v > 0$ by the weaker one: for a set of primes p of density $> 1/2$, $v|p \Rightarrow R_v > |p|_v^{\frac{1}{p-1}}$.

5.4 - <u>Cosingularities of G-operators</u>

In this sub-section we give a <u>corollary</u> of the part of Katz's theorem which concerns the Fuchsian property. Let us make some comments on this property.

i) Let G represent a Fuchsian ∂-module M over $K(x)$. Hence for every $s \in \mathbb{P}^1_{\bar{\mathbb{Q}}}$ there exists $H_s \in GL_\mu(\bar{\mathbb{Q}}((x-s)))$ such that $(x-s)H_s[G] \in M_\mu(\bar{\mathbb{Q}}[[x-s]])$ (with the usual convention $x-\infty = 1/x$). By truncating H_s at sufficiently high order N in $x-s$, one still has $(x-s)H_{s,\leq N}[G] \in M_\mu(\bar{\mathbb{Q}}[[x-s]])$, but also $H_{s,\leq N}[G] \in M_\mu(\bar{\mathbb{Q}}(x))$. This means that for every $s \in \mathbb{P}^1_{\bar{\mathbb{Q}}}$ there exists a $\bar{\mathbb{Q}}[x]$-lattice in $M_{\bar{\mathbb{Q}}}$ stable under $(x-s)d/_{d(x-s)}$.

ii) Let us denote by $\mathrm{Sin}\, L \subset \bar{\mathbb{Q}}$ the set of <u>non-apparent finite singularities</u> of any differential operator $L = \partial - G$ over $K(x)$; let $q_{\sin}$, resp. $q_{a\cos}$, stands for the product $\prod (x-\zeta)$ extended over all elements of $\sin L$, resp. over all apparent finite singularities (which are certain cosingularities, see III 3.2). We put $\Gamma_m = x^{-m} \frac{G_{[m]}}{m!}$.

The classical theory of logarithmic singularities tells us that if M is Fuchsian, then there exists some integer b such that $q^b_{a.\cos}\, q^{m+b}_{\sin}\, \Gamma_m \in M_\mu(\bar{\mathbb{Q}}[x])$, and $\mathrm{ord}_0(\frac{1}{x^m} G_{[m]}(\frac{1}{x})) \geq -m-b$ (condition at ∞), see e.g. [21] II 1. The second assertion may be restricted as:

$\deg(q^b_{a.\cos}\, q^{m+b}_{\sin}\, \Gamma_m) \leq (|\mathrm{Sin}\, L|-1)m+c$ for some integer c independent of m .

iii) Let $\xi \in K$ be a <u>cosingularity</u> of a <u>scalar</u> Fuchsian differential operator Λ . By change of variable $x \longmapsto x-\xi$, we may assume that $\xi=0$, and write Λ in the standard form $\partial - G$, with

$$G = \begin{pmatrix} 0 & 1 & & 0 \\ & \ddots & \ddots & \\ & & & 1 \\ * & * & \cdots & * \end{pmatrix}$$

, see final remark in III 3. Because Λ is scalar and Fuchsian, $G \in M_\mu(K[[x]])$. On the other side, taking into account notations III 1, the fact that 0 is a cosingularity implies that $x^C = X^{-1}Y \in GL_\mu(\bar{\mathbb{Q}}((x)))$. Hence C is diagonalizable with integer-eigenvalues. By shearing along the lines of III 3.1, we get a matrix $H^{sh} \in GL_\mu(K[x,\frac{1}{x}])$ such that $H^{sh}[G](0) = 0$, i.e. 0 is an ordinary point. We turn back to the original variable by the change $x \longmapsto x+\xi$, and get a differential system $\partial - G'$ which represents the same ∂-module as Λ , and has the same set of singularities, except for ξ which disappears.

PROPOSITION. Let Λ be a scalar G-operator, and let ξ be a cosingularity of Λ . Then any $y \in K((x-\xi))$ annihilated by Λ satisfies

$$\rho(y(x+\xi)) \leq \rho(\Lambda) + (|\mathrm{Sin}\Lambda|-1)h_f(\xi) + \sum_{\zeta\in\mathrm{Sin}\Lambda} (h_f(\zeta) + h(\zeta-\xi))$$

$$\leq \kappa_1 h(\xi) + \kappa_2 + \rho(\Lambda) ,$$

where $\kappa_1 = 2|\mathrm{Sin}\Lambda|-1$ and $\kappa_2 = |\sin\Lambda|\log 2 + 2\sum_{\zeta\in\mathrm{Sin}\Lambda} h(\zeta)$.

Furthermore, one can replace the global radius by the size in these inequalities, which also extent to the case of a general G-operator $\partial - G$ if ξ is an apparent singularity. The constants κ_1 and κ_2 are invariant under taking symmetric powers.

This improves on a result of Bombieri [7] lemma 17 (and corrects a misprint there: there should be a factor 2 before $\Sigma h(\zeta)$) .

Proof: remind that $\rho(\Lambda)$ and $\sigma(\Lambda)$ are invariants of the associated $K(x)[\partial]$-module; on the other side it is plain to check that for any $H \in M_\mu(k[x,1/x])$ and $Y \in k((x))^\mu$, one has $\rho(HY) \leq \rho(Y)$ and $\sigma(HY) \leq \sigma(Y)$.
Hence, in the statement one may replace Λ by the G-operator $\partial - G'$ constructed in remark iii) above, which has same set Sin , and for which 0 is ordinary. When ξ is an apparent singularity, the case of a general G-operator of the shape $\partial - G$ can be dealt with along similar lines: a trick due to Birkhoff-Christol (see next chapter, § 1) allows to transform the apparent singularity ξ into an ordinary point by means of a matrix in $GL_\mu(K_v[x-\xi,\frac{1}{x-\xi}])$. Since all invariants are not changed by any finite extension of K , we may and shall start

from $L = \partial - G$ such that ξ is an ordinary point, and such that $\mathrm{Pol}\, G \subset K$. We have to compare $h_{v,n}(y) = \frac{1}{n}\log^+ \mathrm{Max}_{m\leq n} \| \Gamma_m(\xi) \|_v$ and $h_{v,n}(L) = \frac{1}{n}\log^+ \mathrm{Max}_{m\leq n} \| \Gamma_m(t_v) \|_v$, with the notation of remark ii) above.

Since the entries of $q^b_{a.\cos}\, q^{m+b}_{\sin}\, \Gamma_m$ are polynomials of degree $\leq (|\mathrm{Sin} L|-1)m+c$, for some b,c independent of m, one has the bound

$$\| q^b_{a.\cos}(\xi) q^{m+b}_{\sin}(\xi)\Gamma_m(\xi) \|_v \leq \| q^b_{a.\cos}(t_v) q^{m+b}_{\sin}(t_v)\Gamma_m(t_v) \|_v \mathrm{Max}(1,|\xi|_v)^{(|\sin L|-1)m+c},$$

hence (for $v \in \Sigma_f$):

$$h_{v,n}(y) \leq (|\sin L|-1+\tfrac{c}{n})\log^+|\xi|_v + (1+\tfrac{b}{n}) \sum_{\zeta\in \mathrm{Sin} L} \log^+|t_v-\zeta|_v$$

$$+ \frac{b}{n} \sum_{\substack{\zeta\in\mathrm{Pol}\Gamma_1 \\ \zeta\notin \mathrm{Sin}L}} \log^+|t_v-\zeta|_v + (1+\tfrac{b}{n}) \sum_{\zeta\in\mathrm{Sin}L} \log^+ \frac{1}{|\xi-\zeta|_v}$$

$$+ \frac{b}{n} \sum_{\substack{\zeta\in\mathrm{Pol}\Gamma_1 \\ \zeta\notin \mathrm{Sin}L}} \log^+ \frac{1}{|\xi-\zeta|_v} + h_{v,n}(L) .$$

Note that $\log^+|t_v-\zeta|_v = \log^+|\zeta|_v$ for algebraic ζ. Now we have $\rho(y) = \rho_\infty(y) + \rho_f(y)$ and $\sigma(y) \leq \rho_\infty(y) + \sigma_f(y)$, and

$$\rho_\infty(y) = \sum_{\zeta\in\mathrm{Sin}L} \sum_{v\in\Sigma_\infty} \log^+ \frac{1}{|\xi-\zeta|_v} .$$

Using the formulae which relate $\rho_f(y)$, $\sigma_f(y)$, $\rho(L)$ and $\sigma(L)$ to the $h_{v,n}$'s, we reach at last the first required estimate in the proposition. The second one follows, since $h(\frac{1}{\xi-\zeta}) = h(\xi-\zeta) \leq \log 2 + h(\xi) + h(\zeta)$. The invariance of κ_1 and κ_2 under taking $\otimes N$ is left to the reader.

□

The study of solutions at singular points which are not cosingularities requires deeper techniques and will be the main aim of the next chapter.

5.5 - <u>Duality</u>

In this last sub-section, we give another <u>corollary</u> of Katz's theorem, where exponents come into play.

PROPOSITION. <u>Let</u> M <u>be a</u> ∂<u>-module of rank</u> μ <u>over</u> $K(x)$. <u>Then for the dual module</u> M^*, $\sigma(M^*) \leq \sigma(M)(1+\log(\mu-1))$.

Proof: we may assume that $\sigma(M) < \infty$, whence $\rho(M) < \infty$ by theorem 5.2, and then $\rho(\det M) < \infty$ (lemma 2.d of § 3). Let us choose a basis $\{m_i\}$ of M, whence a differential operator $\partial - G$, and let X_{t_v} be a solution in $GL_\mu(A(t_v, R_v))$, normalized by $X_{t_v}(t_v) = I$. We have $h_{v,n}(M^*, \{m_i^*\} = \frac{1}{n}\log^+ \operatorname{Max}_{m \leq n} \| \left({}^t X_{t_v}^{-1}\right)_m \|_v$

$$\leq \frac{1}{n}\log^+ \operatorname{Max}_{m \leq n} \| \left(X_{t_v}^{\otimes \mu-1}\right)_m \|_v + \frac{1}{n}\log^+ \operatorname{Max}_{m \leq n} \left| \left(\det X_{t_v}^{-1}\right)_m \right|_v .$$

Note that $(\det M, \Lambda m_i)$ is represented by $\partial - \operatorname{tr} G$, which annihilates $\det X_{t_v}$. By Katz's theorem, $\partial - \operatorname{tr} G$ is Fuchsian with rational exponents $\alpha_s = \operatorname{Res}_s \operatorname{tr} G (s \in \operatorname{Pol} G)$, $\alpha_0 = \det G(0)$; by Fuchs' condition, $\operatorname{tr} G = \sum \frac{\alpha_s}{x-s} + \alpha_0$, so that $\det X_{t_v} = \left(\frac{(x-t_v)}{t_v} + 1\right)^{\alpha_0} \prod_{s \in \operatorname{Pol} G} \left(1 - \frac{(x-t_v)}{s-t_v}\right)^{\alpha_s}$. Let $V \subset \Sigma_f$ be the finite set of places v such that one of the quantities $|\alpha_0|_v$, $|s|_v$, $|\alpha_s|_v$ is $\neq 1$ (since the proposition is invariant under finite extension of K, we have assumed that the poles s belong to K). Then for any $v \notin V$, we can see on the expansion of the algebraic function $\det X_{t_v}$ at t_v that $1 = |\det X_{t_v}| = |\det X_{t_v}(t_v)| = |\det X_{t_v}^{-1}|$. Using once again Shidlovski's reordering trick, one finds

$$(*)\quad h_{v,n}(M^*, \{m_i^*\}) \leq h_{v,n}(M, \{m_i\})(1 + \log(\mu-1)).$$

On the other hand, we have shown (in the course of proving theorem 5.2) that $\lim_n h_{v,n}$ does exist. This implies the following equality, since V is a finite set:

$$\sigma(M^*) = \sum_{v \in V} \lim_n h_{v,n}(M^*, \{m_i^*\}) + \overline{\lim_n} \sum_{v \in \Sigma_f \setminus V} h_{v,n}(M^*, \{m_i^*\})$$

$$= \sum_{v \in V} \log 1/R_v(M^*) + \overline{\lim_n} \sum_{v \in \Sigma_f \setminus V} h_{v,n}(M^*, \{m_i^*\}) .$$

A similar equality holds for $\sigma(M)$ as well, and we have $R_v(M) = R_v(M^*)$ (prop.1 in 2.5). Putting together this and (*), we find the required inequality.

□

5.6 - Size and local monodromy.

The last part of Katz's theorem may be interpreted as asserting that all local monodromies of a G-operator are quasiunipotent; indeed, the eigenvalues of the local monodromy T at some point $s \in \mathbb{P}^1_{\bar{\mathbb{Q}}}$ are the exponentials of the exponents $x2i\pi$. Now we are going on relating (partially) the size to the level of quasi-unipotency, that is, the minimal positive integer ℓ such that $(T^n-I)^{\ell} = 0$ for $n >> 0$. Let M denote a ∂-module over $K(x)$.

PROPOSITION 1. <u>If</u> $\sigma(M) < \rho(M)+1 < \infty$, <u>then all local monodromies of</u> M <u>are finite</u>.

Since $\sigma(M) < \infty$, M belongs to the Fuchsian class. Let us choose a suitable basis of M in which the connection is expressed by a matrix $G \in M_\mu(K[x]_{(x)})$. We put $L = \partial - G$, and we let N denote the nilpotent part in the additive JOrdan decomposition of $G(0)$: $G(0) = D+N$; D has only rational eigenvalues. The proposition will be derived from the following one:

PROPOSITION 2. <u>If</u> $N \neq 0$, <u>then for any</u> $m > 0$,
$\overline{\lim_n} \sum_{p(v)>m} h_{v,n}(L) \geq 1$.

In order to recover proposition 1, it suffices to show that $\sigma(M)<\rho(M)+1$ implies that $\overline{\lim_n} \sum_{p(v)>m} h_{v,n}(L') < 1$ for some suitable m and any differential operator L' obtained from L by change of variable $x \longmapsto x-\zeta$, $\zeta \in \operatorname{Sing} L$ (possibly ∞). This follows from the following lemma, in which [ii) $n = -1$] is used to handle the case $\zeta = \infty$, and the following argument:

Since $\rho(M) = \sum_{v\in\Sigma_f} \overline{\lim_n} h_{v,n}(L) < \infty$, one still has $\sigma(M) - \sum_{p(v)\leq m} \overline{\lim_n} h_{v,n}(L) < 1$ for sufficiently large m. But we have seen in the proof of theorem 5.2 that for every $v \in \Sigma_f$, $(h_{v,n}(L))$ has a limit; hence $\sigma(M) \leq \overline{\lim_n} \sum_{p(v)>m} h_{v,n}(L) + \sum_{p(v)\leq m} \lim_n h_{v,n}(L)$, and we obtain $\overline{\lim_n} \sum_{p(v)>n} h_{v,n}(L) < 1$.

LEMMA. Let L' denote the differential operator obtained from L after either of the following changes of variable

i) $x \longmapsto x-\zeta$, $|\zeta|_v \leq 1$,

ii) $x \longmapsto x^k$, for a non-zero integer k.

Then $h_{v,n}(L') \leq h_{v,n}(L)$ for any $n \in \mathbb{N}$.

Proof: in case i), this follows from the easy formula $G'_{[n]} = \left(\frac{x}{x-\zeta}\right)^n G_{[n]}(x-\zeta)$. For case ii), one considers the complete solution analytic at t^k (which is another generic point), say X, which satisfies $X(t^k) = I$; one has the following expansions:

$$X(x^k) = \sum \frac{G_{[n]}(t^k)}{n!t^{kn}}(x^k-t^k)^n = \sum \frac{G_{[n]}(t^k)}{n!t^{kn}}(x-t)^n((x-t)^{k-1}+\ldots+kt^{k-1})^n$$

$$= \sum \frac{G'_{[n]}(t)}{n!t^n}(x-t)^n \text{ , whence the result.} \qquad \square$$

Proof of proposition 2: assume that $N \neq 0$ and consider $\overline{\lim_n} \sum_{p(v)>m} h_{v,n}(L)$ for $m >> 0$. By changing the variable $x \longmapsto x^k$ for k = common denominator of the eigenvalues of D, and then shearing, one may assume that $D = -I$; here one uses the previous lemma [ii], and the fact that a rational transformation $\partial - G \longmapsto \partial - H[G]$ does not affect $h_{v,n}$ for almost every v. Moreover, one may assume that m is large enough so that the poles of G are v-adic units in $\mathbb{C}_v$, and that $\|N\|_v = 1$ if $p(v) > m$. In this situation, we have

$$h_{v,n}(L) = \frac{1}{n}\log \max_{\ell \leq n} \left\| \frac{G_{[\ell]}}{\ell!} \right\|_v$$

$$\geq \frac{1}{n}\log \max_{\ell \leq n} \left\| \frac{G_{[\ell]}(0)}{\ell!} \right\|_v$$

$$= \frac{1}{n}\log \max_{\ell \leq n} \left\| \binom{N-I}{n} \right\|_v ,$$

hence

$$\sum_{p(v)\geq m} h_{v,n}(L) \geq \frac{1}{n} \sum_{p\geq m} \log \max_{\ell\leq n} \left| \sum_{k\leq \ell} 1/k \right|_p$$

because $\binom{N-I}{n} = \sum_{j=0}^{n} (-1)^{n-j} \sum_{k_1<\ldots<k_j\leq \ell} (k_1\ldots k_j)^{-1}N^j$, and the non-zero powers of N are linearly independent. But $\operatorname{Max}_{\ell\leq n}\left|\sum_{k\leq\ell} 1/k\right|_p = \operatorname{Max}_{\ell\leq n}|1/\ell|_p$, and we obtain at last

$$\sum_{p(v)\geq m} h_{v,n}(L) \geq \frac{1}{n}\log \text{L.C.M.}(2,3,\ldots,n) - m\,\frac{\log n}{n} .$$

We conclude by invoking once again the prime number theorem. □

REMARK. One gets nothing more in considering the coefficient of other powers of N .

OPEN PROBLEMS. Is $\sigma(M)-\rho(M)$ always an integer? May one interpret it in terms of v-curvatures (see ex.4 below)? Note however that the order of nilpotency of v-curvatures grows linearly when M is replaced by its successive symmetric powers, while $\sigma(M)-\rho(M)$ grows at most in logarithmic rate.
Also, on paraphrasing Grothendieck's conjecture, does the equality $\sigma(M) = \rho(M)$ imply that M is solvable by means of algebraic functions over $K(x)$? (This can be checked when M is "solvable by quadratures" using methods of chapter VIII).

EXERCISES.

1) Prove the statements asserted in remark ii) of § 5.4.

2) Let Λ be an element of $A_1(\mathcal{O}_K) = \mathcal{O}_K[x,d/dx]$. Remark that the residue field of $\mathcal{O}_K$ at v is contained in $\mathbb{F}_{p^{d_v}}$. The v-curvature of Λ , say $\psi_v(\Lambda)$, is by definition the image of $\frac{d^p}{dx^p}$ in $A_1(\mathbb{F}_{p^{d_v}})/A_1(\mathbb{F}_{p^{d_v}})\Lambda$. Compare ψ_v to the differential polynomial denoted by Λ_p in III 4. Show that left multiplication by ψ_v in $A_1(\mathbb{F}_{p^{d_v}})/A_1(\mathbb{F}_{p^{d_v}})\Lambda$ is an endomorphism of $A_1(\mathbb{F}_{p^{d_v}})$ -module.

3) (Cartier) Assume that the order of Λ does not decrease by specializing mod. v . Show that Λ is solvable in $\mathbb{F}_{p^{d_v}}[[x]]$ iff

$\psi_v = 0$. Show that ψ_v is nilpotent iff $\Lambda \otimes 1$ (in $A_1(\mathbb{F}_{p^{d_v}}) = A_1(\mathcal{O}_K) \otimes \mathbb{F}_{p^{d_v}})$ can be expressed as a product of Λ_i's which are solvable in $\mathbb{F}_{p^{d_v}}[[x]]$.

4) Let $G \in M_\mu(K(x))$ such that $\|G\|_v \leq 1$. If G arises from a differential equation as in III 3.2, show that $\psi_v = 0$ iff $\|G_{[p]}\|_v < 1$, and that ψ_v is nilpotent iff $\|G_{[pn]}\|_v < 1$ for some n . In the general case, show that the following statements are equivalent:

i) $\|G_{[pn]}\|_v < 1$ for some n ,

ii) $\|G_{[n]}\|_v < 1$ for every $n \geq p\mu$,

iii) $R_v(\partial - G) > |p|^{1/p-1}$.

5) Give an example of a differential operator Λ which has vanishing p-curvature infinitely many times exactly on a set of primes of density 0. (Hint: consider an elliptic curve E over $\mathbb{Q}$ without complex multiplication. Take for Λ the rank two differential operator which kills the "logarithm" of E , and show that Λ mod p has a full set of solutions in $\mathbb{F}_p[[x]]$ if and only if E has supersingular reduction at p).

6) Let Λ be a G-operator. Show that if $|\mathrm{Sin}\Lambda| < 2$, say $\mathrm{Sin}\Lambda \subset \{1\}$, then Λ is (completely) solvable in terms of polynomials in $L_1 = \Sigma\, x^n/n$ and $(1-x)^r$, $r \in \mathbb{Q}$ (Hint: use the fact that $\pi_1(\mathbb{P}^1 \setminus$two points) is a cyclic group, and see ex.8 of chapter 1).

APPENDIX: OUTLINE OF A THEOREM OF DWORK-ROBBA

This is a deep refinement of Dwork-Frobenius lemma. As explained before, this result is not logically necessary in the present book, se we shall be very brief; see [25] and [7] for more details.

THEOREM. Let $\Lambda = \frac{1}{\mu!}\frac{d^\mu}{dx^\mu} - \sum_{j=0}^{\mu-1}\gamma_j \frac{1}{j!}\frac{d^j}{dx^j}$ be a differential operator with coefficients $\gamma_j \in M(a,r)$. Assume that the kernel of Λ in $M(a,r)$ is of dimension μ , and define Λ_m as in III 4:

$$\Lambda_m = \frac{1}{m!}\frac{d^m}{dx^m} - \sum_{j=0}^{\mu-1}\gamma_{m,j}\frac{1}{j!}\frac{d^j}{dx^j} \in M(a,r)[d/dx]\Lambda .$$

Then we have $|\frac{1}{j!}\gamma_{m,j}|_a(r) \leq \{m,\mu-1\}r^{j-m}$, for every $m \geq 0$ and $0 \leq j \leq \mu-1$, with $\{m,\mu-1\} = \operatorname{Sup}|l_1 \ldots l_{\mu-1}|^{-1}$ (where the sup is over all sets of $\mu-1$ distinct integers bounded by m).

Outline of proof: it is enough to handle the case $a=0$, $r=1$. Let us consider a Frobenius factorization as in 2.1. In order to shorten notation, we set $D = d/dx$, and D^{-1} for its formal inverse.

The proof is based on the following formula, which is checked by induction using Leibnitz's rule:

$$\begin{cases} \frac{1}{m!}D^m y_k = \sum_{j=0}^{k} b_{m,j}(v_0,\ldots,v_j)D^{-1}v_{j+1}D^{-1}\ldots D^{-1}v_k , \text{ with} \\ b_{m,j}(v_0,\ldots,v_j) = \sum \frac{D^{l_0}v_0}{l_0!}\cdots\frac{D^{l_j}v_j}{l_j!}\,\frac{1}{(1+l_j)(2+l_j+l_{j-1})\ldots(j+l_j+\ldots+l_1)} \end{cases}$$

where Σ runs over all $l_i \geq 0$ such that $j + l_j + \ldots + l_0 = m$. Using (1.5.1) one finds $|(v_0\ldots v_j)^{-1}|b_{m,j}|_0(1) \leq \{m,j\}$. On the other hand, we have $y_k = v_0D^{-1}v_1D^{-1}\ldots D^{-1}v_k$, and $\frac{1}{m!}D^n y_k = \sum_{j=0}^{\mu-1}\gamma_{m,j}\frac{1}{j!}D^j y_k$. By putting this into (*), we find

$$\sum_{j=0}^{k} b_{m,j}D^{-1}v_{j+1}D^{-1}\ldots D^{-1}v_k = \sum_{j=0}^{k}\Big(\sum_{h=0}^{\mu-1}\gamma_{m,h}b_{h,j}\Big)D^{-1}v_{j+1}D^{-1}\ldots D^{-1}v_k \text{ for } j,k = 0,\ldots,\mu-1 ,$$

which yields $b_{m,j} = \sum_{h\geq j}^{\mu-1}\gamma_{m,h}b_{h,j}$ for $j = 0,\ldots,\mu-1$. Dividing both sides by $v_0,\ldots,v_j$, one may deduce (using $(v_0,\ldots,v_j)^{-1}b_{j,j} = 1/j!$) :

$$|\tfrac{1}{j!}\gamma_{m,j}|_0(1) = |(v_0,\ldots,v_j)^{-1}(b_{m,j} - \sum_{h=j+1}^{\mu-1}\gamma_{m,h}b_{h,j})|_0(1)$$

$$\leq \max(\{m,j\}, |\gamma_{m,h}|_0(1)\{h,j\} \text{ for } h = j+1,\ldots,\mu-1)$$

$$\leq \{m,\mu-1\}$$

the last step coming by descending induction on $|\frac{1}{j!}\gamma_{m,j}|_0(1)$.

□

Chapter V Local Methods

§ 1. RESOLUTION OF APPARENT SINGULARITIES

We come back to the general p-adic setting of IV 1,2. Thus k denotes an algebraically closed field of characteristic 0 complete under a ultrametric absolute value $|\,|$ with residue field of characteritic $p > 0$, E_0 stands for the complete ring of analytic elements in $D(0,1)$, $E_{00} = E_0 \otimes_{k[x]} k[x]_{(x)}$, etc. ...

Let us consider a matrix $G \in M_\mu(E_{00})$ such that the differential system (1) $\partial X = GX$ has only apparent singularities in $D(0,1)\setminus\{0\}$. This means that for any $a \in D(0,1)\setminus\{0\}$, there exists some $r_a > 0$ and some matrix $X_a \in M_\mu(\,A(a,r_a))$ of rank μ which satisfies (1). The following trick due to Birkhoff-Christol [12] 8.2 allows to get rid of the apparent singularities (= poles of G in $D(0,1)$) in an effective way. It follows from (1) that these singularities (in finite number) correspond to the set of $a \in D(0,1)\setminus\{0\}$ such that $\mathrm{ord}_a \det X_a > 0$ i.e. the set of zeroes of the Wronskian determinant. For such a zero, say a, of $\det X_a$, there exists a linear dependence relation over k, say $\Sigma b_j\,{}_{ij}X_a(a) = 0$, $i = 1,\dots,\mu$, between the columns of X_a; denoting by i_a an index such that $|b_{i_a}| = \mathrm{Max}|b_i|$, one may suppose in addition that $b_{i_a} = 1$. Let us denote by H_a the $\mu \times \mu$ matrix:

$$H_a = \begin{pmatrix} 1 & & & & & & & & \\ & \ddots & & & & \mathbf{0} & & & \\ & \mathbf{0} & 1 & & & & & & \\ -b_1, & \ldots, & -b_{i_a-1}, & x-a, & -b_{i_a+1}, & \ldots, & -b_\mu \\ & & & & 1 & & \mathbf{0} & \\ & & \mathbf{0} & & & \ddots & & \\ & & & & & & 1 \end{pmatrix},$$

so that ${}_{i_a i_a}(H_a) = x-a$, and

$$H_a^{-1} = \begin{pmatrix} 1 & & & & & & & \\ & \ddots & & & \mathbf{0} & & & \\ & \mathbf{0} & 1 & & & & & \\ \frac{b_1}{x-a}, & \ldots, & \frac{b_{i_a-1}}{x-a}, & \frac{1}{x-a}, & \frac{b_{i_a+1}}{x-a}, & \ldots, & \frac{b_\mu}{x-a} \\ & & & & 1 & & \mathbf{0} \\ & & \mathbf{0} & & & \ddots & \\ & & & & & & 1 \end{pmatrix} \longleftarrow \text{row } i_a .$$

Hence $H_a \in GL_\mu(k[x-a,\frac{1}{x-a}])$. Moreover, the relation between columns of X_a show that $X_a' := H_a^{-1}X_a$ still belongs to $M_\mu(A(a,r_a))$, and $\mathrm{ord}_a \det X_a' = \mathrm{ord}_a \det X_a - 1$. By multiplying such matrices H_a , i.e. by setting $H^{res} = \prod H_a$ where the product extends over all zeroes of "the" Wronskian determinant (counted with multiplicity), one can reach the situation where $(H^{res})^{-1} X_a \in GL_\mu(A(a,r_a))$ for any $a \in D(0,1)\setminus\{0\}$; hence $H[G]$ has no pole at all in $D(0,1)$. Using the explicit form of H_a and making use of remark IV 2.2, we see that the following lemma holds true.

LEMMA. *Let* $G \in M_\mu(E_{00})$ *such that the non-zero singularities of the system* $\partial X = GX$ *in* $D(0,1)$ *are apparent ones. Then there exists* $H^{res} \in M_\mu(k[x]) \cap GL_\mu(k[x]_{(x)})$ *such that*

a) $H^{res}[G] \in M_\mu(E_0)$

b) $\| H^{res} \| = \| H^{res-1} \| = 1$

c) $\| H^{res}[G] \| \leq \text{Max}(1, \| G \|)$

d) $H^{res}(0) \in GL_\mu(k)$.

Proof: straightforward, from what precedes.

□

REMARK 1. We shall sometimes encounter the dual situation, where $G \in M_\mu(E_{00})$ and the non-zero singularities of $\partial + {}^tG$ are apparent. We then construct H^{res*} corresponding to the resolution of apparent singularities for this dual system, and we set $H^{cores} := {}^t(H^{res*})^{-1}$. Thus $H^{cores} \in GL_\mu(E_{00})$, $H^{cores-1} \in M_\mu(E_0)$ and $H^{cores}[G] \in M_\mu(E_0)$. Moreover the properties b) c) d) carry over for H^{cores} . Let us also note that they imply $\| H^{cores-1}(0) \| \leq 1$.

REMARK 2. When $G \in M_\mu(\bar{\mathbb{Q}}(x))$, for K a number field, a similar construction works globally, using the embedding $\bar{\mathbb{Q}}(x) \subset \bar{\mathbb{Q}}((x-a))$; one gets $H^{res} \in GL_\mu(\bar{\mathbb{Q}}(x))$.

§ 2. ANALYTIC FROBENIUS FUNCTOR

2.1 - At this point, the reader should remember the setting of III 5; we considered there a functor Φ on ∂-modules which consists essentially in making the change of variable $x \longmapsto x^p$ (and twisting the "coefficients" by an automorphism φ of k). We were looking for an inverse, and found the functor Ξ which inverts Φ up to some twists $[i/p]$ if the base ring A is a free $\varphi(A)$-module generated by $1, x, \ldots, x^{p-1}$. Here we shall specialize this setting by choosing $A = E_0$, E_{00} or else E , and p = residual characteristic
We assume in addition to the previous hypotheses on k that :
(2.1.1) there exists a continuous automorphism of k , denoted by φ , whose restriction to the Witt ring of the residual field is the Frobenius automorphism (see [53]II 6).
Note that φ is necessarily an isometry. One can check that

(2.1.1) (without the hypothesis that k is algebraically closed) is stable under finite extension of k; hence it holds for $\bar{\mathbb{Q}}_p$, and therefore for $\mathbb{C}_p$ as well (by continuity).
The endomorphism ϕ on $k[x]$, defined by $\phi(f) = f^{\varphi}(x^p)$, extends to an isometry of $(k(x),|\,|)$, and of E henceforth by (uniform) continuity. For any $G \in M_\mu(E)$, $\|G^{\phi}\| = \|G\|$; hence ϕ is injective in $M_\mu(E)$. Also, formula (27) of III.5 goes through. On $k[[x]]$, radii of convergence behave well under ϕ: $R(f) = R(\phi(f)))^p$.

LEMMA. <u>For</u> $A = E_0, E_{00}$, <u>or</u> E, A <u>is a free</u> $\phi(A)$-<u>module generated by</u> $1, x, \ldots, x^{p-1}$.

Proof: this would be quite clear for $A = k(x)$; this means that $k(x) \subset k((x))$ is stable under the mappings $\psi_i : \Sigma f_n x^n \longrightarrow \Sigma f^{\varphi^{-1}}_{i+np} x^n$. From (29): $f = \sum_{i=0}^{p-1} x^i \phi\psi_i f$, it follows that

i) $\psi_i(k[x]_{(x)}) \subset k[x]_{(x)}$ (clear), and

ii) ψ is a contraction of $(k(x),|\,|)$. Indeed, let f be a non-zero rational function. Since $\psi_i f$ are also rational functions, not all 0, the existence of $a \in k$ such that $|a| = (\operatorname{Max}_{i=0,\ldots,p-1} |\psi_i f|)^{-1}$ is clear from the very definition of the Gauss absolute value. Replacing f by $a\varphi f$ (making use of the formulae $\psi_i(a\varphi f) = a\psi_i(f)$ and $|a| = |a^{\varphi}|$), we are reduced to prove that $|f| = 1$ if $\operatorname{Max}|\psi_i f| = 1$. But this follows from (29) modulo the valuation ideal of $(k(x),|\,|)$. By uniform continuity, the maps ψ_i extend to the completion E of $k(x)$, thus proving the lemma for $A = E$. On the other hand, one checks readily on the definition of ψ_i that these maps stabilize the subring $A(0,1)$ of $k((x))$ (using Hadamard's formula). This stability, combined with point ii), yields $\psi_i(E_0) \subset E_0$, thus proving the lemma for $A = E_0$; combining further with point i), and reminding that $E_{00} = E_0 \otimes_{k[x]} k[x]_{(x)}$, we get the required assertion for $A = E_{00}$.

□

2.2 - LEMMA 1. <u>Let</u> N <u>be a</u> ∂-<u>module over</u> E_0, <u>solvable in</u> $A(t^p,|p|)$. <u>Then its exponents</u> α_i (at 0, see III 2.3) <u>satisfy</u>

$|\alpha_i| < |p|^{-1}$.

Proof: let F be a representative matrix in $M_\mu(E_0)$ for N ; by conjugating by a constant invertible matrix, we may assume that $F(0)$ is triangular. Using III(9), one gets

$$\left\| \frac{F^{(n)}}{n!} \right\| \geq \left\| \frac{F^{(n)}(0)}{n!} \right\| = \left\| \binom{F(0)}{n} \right\| \geq (\operatorname{Max}|\alpha_i|)^n |n!|^{-1} \quad \text{if}$$

$\operatorname{Max}|\alpha_i| > 1$, hence

$$\operatorname{Max}|\alpha_i| \leq \varlimsup_{n\to\infty} |n!|^{1/n} \left\| \frac{F^{(n)}}{n!} \right\|^{1/n} \leq |p|^{\frac{1}{p-1}} \cdot R(N)^{-1} < |p|^{-1} .$$

□

2.3 - Until the end of § 4, we shall <u>fix a real number</u> R satisfying

(2.3.1) $|p|^{1/p} < R \leq 1$.

Remember that the functor Φ_a is defined using the semi-linear change of variable $\phi_a : x - a \longmapsto (x-a)^p$, so that $\phi_a(x) = (x-a)^p + a^\varphi$. We have set $\Phi := \Phi_0$. Note the obvious implication $a^\varphi = a^p \Rightarrow |a| = 1$ or $a = 0$. The particular choice we have made for φ is justified by the next lemma.

LEMMA (Christol). <u>Let</u> $a \in k$ <u>such that</u> $a^\varphi = a^p$. <u>Let</u> N <u>be a</u> ∂-<u>module over</u> E_{aa} <u>solvable in</u> $A(t^p,R^p)$ (<u>and hence in</u> $A(a,r)$ <u>for some</u> $r > 0$). <u>Then</u> $N^{\Phi_a} \otimes E \simeq (N \otimes E)^\Phi$.

Proof: We may assume that $a \neq 0$. Let $F \in M_\mu(E_{aa})$ be a representative matrix for N , and let $X \in GL_\mu(A(a,r))$ a solution of $\partial X = FX$ with $X(a) = I$; such a solution exists for r small enough (e.g. $r = \|F\|\,|p|^{1/p-1}$) because a is an ordinary point for this differential system: it is given by

$$X = \sum_{n\geq 0} \frac{F_{[n]}}{n!}(a) \frac{(x-a)^n}{a^n} , \quad X^{-1} = \sum_{n\geq 0} \frac{{}^t(-({}^tF)_{[n]})}{n!}(a) \frac{(x-a)^n}{a^n} .$$

Let us put $X^{\phi_a} = HX^\phi$, so that

$$H = X^\varphi((x-a)^p + a^\varphi) X^\varphi(x^p)^{-1} = \sum_{n\geq 0} \frac{F_{[n]}{}^\phi}{n!} \frac{[(x-a)^p + a^\varphi - x^p]^n}{x^{pn}} .$$

(Since $a^\varphi = a^p$, we get $H \in M_\mu(A(a^p,r)))$. In fact $(x-a)^p + a^\varphi - x^p = pxf$ with $f \in k[x]$, $|f| \leq |a| = 1$. Thus the general term of the series which expresses H may be written

$$h_n := \frac{F_{[n]}(x)^{\Phi}}{n!}\left(\frac{pxf}{x^p}\right)^n \ , \text{ with } h_n \in M_\mu(E) \ ;$$

$\| h_n \| \leq \left(\| \frac{F_{[n]}}{n!} \|^{1/n} |p| \right)^n \longrightarrow 0$ since $R^p > |p|$. Because E is complete, we get $H \in M_\mu(E)$. Using duality, we may write down a similar expansion for H^{-1} , which yields finally $H \in GL_\mu(E)$. This establishes the required isomorphism.

□

2.4 - LEMMA. <u>Let</u> $\mathcal{N}$ <u>be a ∂-module over</u> E <u>solvable in</u> $\mathcal{A}(t^p,R^p)$. <u>Then</u> $\mathcal{M} = \mathcal{N}^{\Phi}$ <u>is solvable in</u> $\mathcal{A}(t,R)$.

Proof: since $R^p > |p|$, we have $|x^p-t^p| > R^p \Rightarrow |x^p-t^p| = |x-t|^p$. The lemma follows at once.

□

§ 3. INVERSION OF THE FROBENIUS FUNCTOR

3.1 - The aim of this paragraph is to describe a sub-functor Ψ of the Christol functor Ξ , which inverts (up to $\otimes E_{00}$) the Frobenius functor Φ restricted to normalized $E_0[\partial]$-modules, solvable in $\mathcal{A}(t,R)$. This construction, due again to G. Christol [12] [13], consists essentially in finding a change of basis such that the new system is expressed in the variable x^p . Unfortunately the results of Christol were not cast in sufficient generality for our purposes (he handles only the case $R = 1$). The new features of the "general" case $R > |p|^{1/p}$ are described by the occurence of apparent singularities, and theorem 3.5 below.

Let $\mathcal{M}$ be a ∂-module over E_0 solvable in $\mathcal{A}(t,R)$, and let $G \in M_\mu(E_0)$ be a representative matrix. Since $R > |p|^{1/p-1}$, we have $\| G_{[n]} \| \to 0$. Following Christol, we introduce the following matrix:

(3.2.1) $\quad H := 1/p \sum_{n\geq 0} (\xi-1)^n/n!\, G_{[n]}$.

Since $1/p \sum_{\xi^p=1} (\xi-1)^n/n! \in \mathbb{Z}_p$ for every integer n , H belongs to the Banach algebra $M_\mu(E_0)$. Moreover, proposition 2.3 of chapter IV (applied for $a = t,\ r \to 1$), yields the bound

(3.2.2) $\quad \|H\| \leq \operatorname{Max}(1, \|G\|^{\mu-1})$.

3.3 - We make in addition the following hypothesis:

(3.3.1) <u>The exponents of</u> $\mathcal{M}$ <u>belong to</u> $p\mathbb{Z}_p$ <u>and satisfy condition</u> (4) of ch. III (i.e. $\mathcal{M}$ is normalized).

Using III, formula (10), we get

$$\begin{aligned} HY_G &= \frac{1}{p} \sum_{\xi^p=1} \sum_{n\geq 0} \sum_{m\geq 0} Y_m \binom{G(0)+m}{n} x^m (\xi-1)^n \\ &= \sum_{m\geq 0} Y_m x^m \sum_{\xi^p=1} \frac{1}{p}\xi^{m+G(0)} \\ &= \Sigma\, Y_{mp} x^{mp} \quad \text{because of (3.3.1)} \\ &= (\psi_0 Y_G)^{\phi} . \end{aligned}$$

A fortiori $H(0) = I$, whence the existence of $H^{inv} := H^{-1}$ in $M_\mu(E_{00})$. Let us set

(3.3.2) $F := (\partial(\psi_0 Y_G) + \frac{1}{p}(\psi_0 Y_G) G(0)\varphi^{-1})(\psi_0 Y_G)^{-1} \in M_\mu(k[[x]])$.

We get:

(3.3.3) a) $H^{inv} \in M_\mu(E_{00})$, $(H^{inv})^{-1} \in M_\mu(E_0)$, $H^{inv}(0) = I$,

b) $H^{inv}[G] = pF^\phi$, and the non-zero singularities of $\partial - pF^\phi$ are apparent,

c) $F(0) = \frac{1}{p} G(0)^{\varphi^{-1}}$

d) $Y_F = \psi_0 Y_G$

e) $F \in M_\mu(E_{00})$, and the non-zero singularities of $\partial - F$ are apparent.

Indeed a) and d) are clear, and d) $\Rightarrow Y_F{}^\phi = HY_G \Rightarrow$ b) $\Rightarrow$ c). It remains to prove e). By b), we know that F^ϕ lies in the subring E_{00} of $k[[x]]$, and thanks to lemma 2.1, it follows that $F = (F^\phi)^{\psi_0} \in M_\mu(E_{00})$ too. Let a^p denote a possible pole of F , and let θ_a be the function $x \longmapsto a(1+x/_a p)^{1/p}$ which is analytic in $D(a^p, |p|^{p/p-1})$; let X_{a^φ} be a solution of $\partial X = pF^\phi X$, analytic at a^φ ; then the matrix $X'_{a^p} = (X_{a^\varphi})^{\varphi^{-1}} \circ \theta_a(x-a^p)$ is a solution of $\partial X' = FX'$, analytic at a^p . This completes the proof of e).

REMARK. If $\|G\| \leq 1$, one deduces $\|G_{[n]}\| \leq 1$ for all n, and $\|H\| \leq 1$. It follows that $|\det H| = |\det H(0)| = 1$, hence $H^{inv} \in GL_\mu(E_0)$, and $F \in M_\mu(E_0)$.

3.4 - We associate to M a ∂-module M^Ψ over E_{00} defined by

$$\begin{cases} M^\Psi = M \otimes_{E_0} E_{00} \text{ as } E_{00}\text{-module} \\ \text{the action of } \partial \text{ on } M^\Psi \text{ is represented by } F \text{ in the basis} \\ \text{where the action of } \partial \text{ on } M \text{ is represented by } G . \end{cases}$$

PROPOSITION (Existence of a weak Frobenius structure). Let M be a normalized ∂-module over E_0, solvable in $A(t,R)$, whose exponents belong to $p\mathbb{Z}_p$. Then $(M^\Psi)^\Phi \simeq M \otimes_{E_0} E_{00}$.

Conversely, let N be a normalized ∂-module over E_0, solvable in $A(t^p,R^p)$, whose exponents belong to $\mathbb{Z}_p$. Then $(N^\Phi)^\Psi \simeq N \otimes_{E_0} E_{00}$.

Proof: the first assertion is a straightforward consequence of (3.3.3) above. Let us deduce the latter from the former one. Lemma 2.1 for $A = E_{00}$ allows to apply proposition 5.3 of ch. III:

$$((N^\Phi)^\Psi)^\Phi \simeq N^\Phi \otimes_{E_0} E_{00} \Rightarrow (((N^\Phi)^\Psi)^\Phi)^\Xi \simeq ((N \otimes_{E_0} E_{00})^\Phi)^\Xi$$

$$\Rightarrow \bigoplus_{i=0}^{p-1} \underbrace{(N^\Phi)^\Psi \otimes [i/p]}_{N'_i} \simeq \bigoplus_{i=0}^{p-1} \underbrace{N \otimes_{E_0} E_{00} \otimes [i/p]}_{N_i} .$$

In order to get the required isomorphism $N'_0 \simeq N_0$, we have to overcome the difficulty raised by the fact that the N'_i and N_i's are not Artinian.

Let us first note that property (1.3.1) in IV implies that E_{00} is a discrete valuation ring with uniformizing parameter x. It follows that for any integer $N > 0$, the $E_{00}[\partial]/_{(x^N)}$-modules $N'_{i,N} := N'_i \otimes_{k[x]} k[x]/_{(x^N)}$ and $N_{i,N}$ (defined in a similar way) are both Artinian and Noetherian. Thereby, we can apply the Krull-Schmidt theorem which asserts existence and unicity (up to permutation isomorphism of the factors) of a decomposition into indecomposable summands [36].

Because $\operatorname{End} N_i = \varprojlim_N \operatorname{End} N_{i,N}$ as ∂-modules (same for N_i'), we also have $\varprojlim_N \operatorname{End}_\partial N_{i,N} = \operatorname{End}_\partial N_i$ (same for N_i'), which are finite-dimensional over k (dim $\leq \mu^2$). Hence $\operatorname{End}_\partial N_i = \operatorname{End}_\partial N_{i,N}$ for $N >> 0$ (same for N_i'). This implies that for $N >> 0$, the decomposition of $N_{i,N}$, resp. $N'_{i,N}$, into indecomposable summands, which is the same as a set of orthogonal minimal idempotents, lifts to N_i, resp. N_i'. So do permutation isomorphisms of direct summands.
Now any direct summand of N_i or N_i' has all its exponents in $i/p + \mathbb{Z}_p$; because exponents are invariant under isomorphism over E_{00}, we get $N_i \simeq N_i'$, and therefore $(N^\Phi)^\Psi \simeq N \otimes_{E_0} E_{00}$.

□

REMARK. Here are some words to justify the notation M^ψ, i.e. to prove that this ∂-module does not depend on the auxiliary basis chosen for the construction. With self-understandable notations, it suffices to prove that

$$ {}^Y\!\left(H^{inv}_{H_1[G]} \cdot H_1\right)[G] = {}^Y\!\left(H_1 \cdot H^{inv}_G\right)[G] \text{ for any } H_1 \in GL_\mu(E_0) . $$

But using the fact that ψ_0 is a morphism of <u>rings</u> $\phi(k[[x]]) \longrightarrow k[[x]]$, an elementary calculation involving formulae (3.3.3)d and III(18) shows that both members are equal to $\psi_0(H_1 \cdot Y \cdot H_1(0)^{-1})$.

3.5 - THEOREM. <u>Let</u> M <u>be a normalized</u> ∂-<u>module over</u> E_0, <u>solvable in</u> $A(t,R)$, <u>whose exponents belong to</u> $p\mathbb{Z}_p$. <u>Then</u> M^ψ <u>is solvable in</u> $A(t^p,R^p)$; <u>in fact</u> $R(M^\psi) = R(M)^p$.

We remind the reader that $R > |p|^{1/p}$.

Proof: Let us choose a representative matrix $G \in M_\mu(E_0)$ for M. Replacing E_0 by E_{00}, we may assume in this proof that $pF^\phi = G$, after a change of basis through H^{inv}. Using (3.3.3)e) and the resolution of apparent singularities, one may assume, in point of fact, that $F \in M_\mu(E_0)$.

i) We handle at first the case when 0 is an ordinary point, i.e. $G(0) = 0$. By (3.3.2) 0 is also an ordinary point for $\partial - F$.

It follows that the expansions of G and F look like this:

$$\frac{t^{-n}G_{[n]}(t)}{n!} = \sum_{m\geq 0} G_{n,m}t^m, \quad \frac{t^{-n}F_{[n]}(t)}{n!} = \sum_{m\geq 0} F_{n,m}t^m \text{ ,where}$$

$F_{n,m}$, $G_{n,m} \in M_\mu(k)$. Here, we have considered the generic point as a variable over k and we have replaced the usual embedding $E_0 \subset k[[x]]$ by $E_0 \subset k[[t]]$; the point is that the expansions start from $m = 0$. Let us next write out an expansion $(x^p-t^p)^n = \sum_{j=0}^{np} \Lambda_{np,j}(x-t)^{np-j}t^j$, and let us complete the set of coefficients, in order to have $\Lambda_{i,j}\in\mathbf{Z}$, $\Lambda_{i,j}=1$ if $i\notin_p\mathbf{Z}$ or $j=0$. Let V , resp. W , be a matrix whose entries are analytic at t (i.e. in a small disk of Ω centered at t), satisfying $\partial V = pF^\phi V = GV$, $V(t) = I$, resp. $\partial W = FW$, $W(t) = I$. Hence we have $W^\phi = V$, and this equality can be translated into

$$(3.5.1) \quad \sum_{n\geq 0} \frac{1}{n!} t^{-pn}(F_{[n]}(t))^\phi(x^p-t^p)^n = \sum_{n\geq 0} \frac{1}{n!} t^{-n}G_{[n]}(t)(x-t)^n$$

To go further, we must assume that φ is extended to Ω in such a way that $t^\varphi = t^p$. In (3.5.1), x and t may be viewed as two independent variables, since Ω is linearly disjoint from $k((x))$ over k .

It is convenient to set

$$H_{n,m} = \begin{cases} F^\varphi_{n/p,m/p} & \text{if } m \text{ and } n \text{ are both divisible by } p \\ 0 & \text{otherwise.} \end{cases}$$

Hence (3.5.1) can be written

$$\sum_{n\geq 0}\sum_{m\geq 0}\sum_{0\leq l\leq n} H_{n,m}\Lambda_{n,l}t^{m+l}(x-t)^{n-l} = \sum_{j\geq 0}\sum_{i\geq 0} G_{j,i}t^i(x-t)^j \text{ , in}$$

$k[[t,x-t]]$. Equating coefficients, we reach the following strict upper triangular linear system:

$$(3.5.2) \quad G_{j,Mp-j} = \sum_{j\leq n\leq M_p} \Lambda_{n,n-j}H_{n,Mp-n} \quad \text{for } j = 0,\dots,Mp ,$$

which can be inverted:

$$(3.5.3) \quad F^\varphi_{n,m} = \sum_{0\leq j\leq np} M_{m,n,j}G_{j,(m+n)p-j} \text{ , where } M_{m,n,j} \in \mathbf{Z} .$$

this yields the bound (since φ is an isometry):

$$\operatorname*{Max}_{m\leq n} \left\|\frac{F_{[m]}}{m!}\right\| \leq \operatorname*{Max}_{j\leq np} \left\|\frac{G_{[j]}}{j!}\right\| \text{ , which implies } r(W) \geq r(V)^p ,$$

i.e. $R(M^{\psi}) \geq R(M)^p$; the reversed inequality follows from (3.5.2).

ii) In the general case, one may choose a residue class $D(a,1)$ in which G extends analytically à la Krasner (see IV 1.3.2): all but a finite number of residue classes are available. We have $|x^{p^{\varphi^{-1}}} - y^{p^{\varphi^{-1}}}| = |x-y|\,|((x-y)^{p-1}+p\ldots)|$, so that $x \longmapsto x^{p^{\varphi^{-1}}}$ is a strict contraction of any residue class $D(a,1^-)$. Hence this contraction has a unique fixed point, which we may take as the center "a". Hence, in the sequel of this proof, we shall assume that $D(a,1)$ is an ordinary disk, with $a^{\varphi} = a^p$. The theory of 3.3 carries over, if we replace 0 by a , ϕ by ϕ_a etc. ... and we get ψ_a which transforms certain $(x-a)d/dx$-modules over E_a into $(x-a)d/dx$-modules over E_{aa} . We denote by N_a the $(x-a)d/dx$-module over E_{aa} constructed from M in this way, but viewed as a ∂-module over E_{aa} (recall that a is ordinary for N_a). By construction, $N_a^{\phi_a} \otimes_{E_{aa}} E \simeq M \otimes_{E_0} E$. Making use of lemma 2.3, we get

$$(N_a \otimes_{E_{aa}} E)^{\phi} \simeq (M^{\psi} \otimes_{E_{00}} E)^{\phi} .$$

Applying the functor Ξ , and using lemma 2.1 and then proposition III 5.3, we obtain

$$\bigoplus_{i=0}^{p-1} [i/p] \otimes N_a \otimes E \simeq \bigoplus_{i=0}^{p-1} [i/p] \otimes M^{\psi} \otimes E .$$

Since exponents are not defined for $E[\partial]$-modules, we cannot use the argument given in the last proposition. Instead, consider the $E_{00}[\partial]$-module $L := \bigoplus_{i=1}^{p-1} [i/p] \otimes M^{\psi}$. The exponents (at 0) of each direct summand of L belong to $\frac{1}{p}\mathbf{Z}_p \setminus \mathbf{Z}_p$; it follows that no factor of L is solvable in $A(t^p, R^p)$ since $R^p > |p|$ and $F \in M_{\mu}(E_0)$ (see lemma 2.2). Using the unicity (up to permutation isomorphism) of the decomposition of the Artinian and Noetherian $E[\partial]$-module $L \otimes E$ into indecomposable factors, we get an embedding

$L \otimes E \hookrightarrow \bigoplus_{i=1}^{p-1} [i/p] \otimes N_a \otimes E$, which is an isomorphism for dimension reasons. The same unicity argument shows that the complementary modules $N_a \otimes E$ and $M^{\psi} \otimes E$ are isomorphic. We get thereby that $R(M^{\psi}) = R(N_a)$, which is also $R(M)^p$ according to the point i) in the proof.

□

REMARK. The identity (3.5.1) does not imply formally $r(W) \geq r(V)^p$. By way of counterexample, take $V = 1 + \frac{x-t}{t}$, $W = (1+(x-t)/t)^{1/p} = \sum_{n\geq 0} (1/p)_n t^{-n} \frac{(t-x)^n}{n!}$, so that $R(W) = |p|^{p/p-1}$.

§ 4. CONVERGENCE OF THE UNIFORM PART

4.1 - Let us consider a differential system $\partial X = GX$ such that

i) $G \in M_{\mu}(E_0)$,

ii) the eigenvalues α_i of $G(0)$ are rational numbers, with common denominator N ,

iii) none of the differences $\alpha_i - \alpha_j$ is a non-zero integer.

LEMMA. Under i), ii), iii), the nth coefficient of the normalized uniform part Y_G of the solution X satisfies:

$$\frac{1}{n}\log^{+}\| Y_{G,n}\| \leq \log^{+}\| G\| + \frac{\mu^2 \log|p|^{-1}}{p-1} + (\mu^2-1)\log^{+}\| G(0)\| \sum_{i,j=0}^{\mu-1} \frac{\log^{+}(\alpha_i - \alpha_j + n)N}{n} .$$

Proof: formula III(13) yields the bound

$$\log\| Y_{G,n}\| \leq \log^{+}\|G\| + (\mu^2-1)\log^{+}\| G(0)\| + \log \Big| \prod_{i,j}(\alpha_i + \alpha_j + n)^{-1}\Big| + $$

$$+ \log \operatorname{Max}_{m<n} \| Y_{G,m}\| .$$

The problem is to estimate $\left| \prod_{m=1}^{n}(\alpha_i - \alpha_j + n)^{-1}\right|$: if $\alpha_i - \alpha_j$ is p-integral we use I, Appendix; otherwise $\left| \prod_{m=1}^{n}(\alpha_i - \alpha_j + n)^{-1}\right| < 1$, and we conclude by induction.

□

4.2 - This inequality implies at once that

$$(4.2.1)\quad R(Y_G) \geq |p|^{\mu^2/p-1}\ \mathrm{Max}(1, \|G\|)^{-1}\mathrm{Max}(1, \|G(0)\|)^{1-\mu^2}\ .$$

However when $\partial X = GX$ is solvable in some generic disk, we shall sharpen this poor estimate.

PROPOSITION. <u>If in addition to the assumptions of the lemma, we assume that</u> $\partial X = GX$ <u>is solvable in</u> $GL_\mu(\,A(t,R))$ <u>with</u> $|p|^{1/p} < R \leq 1$, <u>then</u> $R(Y_G, Y_G^{-1}) \geq R^{(\mu-1)\mu^2}|p|^{\mu^2/p(p-1)}$.

Proof: a change of basis over $E_0[\frac{1}{x}]$ does not change $R(Y_G)$, provided it does not destroy conditions i) and iii) above (see the discussion which follows III(17)). Because $R > |p|$ and the exponents are rational, they belong to $\mathbf{Z}_p$ (lemma 2.2). By shearing, we may change G so that the eigenvalues of $G(0)$ go into $p\mathbf{Z}_p \cap \mathbf{Q}$ and satisfy iii). By means of H^{inv} , we may change G (without any further change of its value at 0) so that G can be written pF^{ϕ} for some $F \in M_\mu(E_{00})$; since only apparent singularities appear through this process, this does not change $R(Y)$ anyway.
Hence we have $R(Y_F) = R(Y_F^{\phi})^p = R(Y_{pF^{\phi}})^p = R(Y_G)^p$. Now we can remove the non-zero singularities of F (which are all apparent, see 3.3.3) by means of H^{res} . Theorem 3.5 above shows that $\partial - F' := \partial - H^{res}[F]$ is solvable in $A(t^p, R^p)$; by a constant change of basis, we may also assume that $F'(0)$ has Jordan normal form. We then choose an ordinary point $b \in D(0,1)$ and transform the system into a differential equation (by means of H^{eq} centered at b firstly, and then put in the standard form

$$F'' = \begin{pmatrix} 0 & 1_1 & & \bigcirc \\ & \ddots & \ddots & \\ \bigcirc & & \ddots & \cdot 1 \\ ** & \cdots & \cdots & * \end{pmatrix}$$

) . According to Dwork-Frobenius lemma (IV 2.1, applied to $a = t^p$ and $r = R$), one has the bound $\|F''\| \leq R^{(1-\mu)p}$. Since 0 is a logarithmic singularity of F'' , Fuchs's criterium (see remark III 3.2) tells that $F'' \in \mathbf{M}_\mu(E_{00})$. Moreover, we know that the poles of F'' are only apparent singularities of $\partial - F''$. Let us remove these singularities by applying a transformation $H^{res'}$ once again. Using lemma 1b), we see that $F''' := H^{res'}[F'']$ still satisfies the inequality $\|F'''\| \leq R^{(1-\mu)p}$. It is not clear whether $\partial - F'''$ is normalized or not; however there does exist a solution of the form $Y'''x^C$ with $Y''' \in GL_\mu(k((x)))$, and with $C = F'(0)$: it suffices to look at the formula III(17) which relates $X_{F'}$ to $X_{F'''}$. The estimate of Lemma 4.1 still holds up to $o(n)$ if we work on the general formula III(12) rather than III(13) (for $n >> 0$, $U(C+nI,F'''(0))$ is invertible), which yields, when applied to Y''' and ${}^tY'''^{-1}$ simultaneously (using duality):

$$R(Y''',Y'''^{-1}) \geq |p|^{\mu^2/p-1}(\mathrm{Max}(1,\|F'''\|))^{-1}(\mathrm{Max}\|C\|,\|F'''(0)\|,1)^{1-\mu^2}$$

where $R(Y''',Y'''^{-1})$ stands for the exterior radius of the annulus of convergence. Since F' was in Jordan normal form, we get

$$R(Y''',Y'''^{-1}) \geq |p|^{\mu^2/p-1}(\mathrm{Max}(1,\|F'''\|))^{-\mu^2} \geq |p|^{\mu^2/p-1}R^{\mu^2(1-\mu)p} ;$$

since $R(Y_G,Y_G^{-1}) = R(Y''',Y'''^{-1})$ by the preceding discussion, we get the required lower bound for $R(Y_G,Y_G^{-1})$.

□

REMARK. For $R = 1$, Christol [13] proves that $R(Y_G,Y_G^{-1}) \geq 1$.

4.3 - MOTIVATION. At this point, we can give our motivation for introducing weak Frobenius structures in G-function theory. The point is that, before applying ψ, we can lay down only poor estimates such as lemma 4.1, which involve some term $\frac{\log p}{p-1}$, i.e. the general term of a divergent series in p. However after applying ψ, we get estimates like the bound for $\log R(Y_G)^{-1}$ in the previous proposition, which involves a summable term $\frac{\log p}{p(p-1)}$: $\sum_{p \text{ prime}} \frac{\log p}{p(p-1)} < 1$. In the sequel we shall use twice more this device.

4.4 - REMARK. We should add a few words concerning the convergence of the uniform part in the Archimedean case $(k=\mathbb{C})$. A famous result due to Frobenius asserts that the normalized uniform part at a logarithmic singularity has non-zero radius of convergence (see e.g. [31]). By Cauchy's theorem, it follows that this radius is exactly the distance from 0 to the nearest non-apparent singularity.

§ 5. FROM $\rho(\Lambda)$ TO $\rho(Y)$

We come back to the arithmetic setting, where K denotes a number field, etc. ... Let $\Lambda = d/dx - \Gamma$ be a differential operator with $\Gamma \in M_\mu(K(x))$, such that a solution at 0 can be expressed by Yx^C , with $Y \in GL_\mu(K((x)))$ and $C \in M_\mu(K)$. We shall use the notation $\mathrm{Sin}\Lambda$ from the previous chapter (set of finite non-apparent singularities).

THEOREM. The following inequality holds:

$$\rho(Y) \leqq \mu^3\rho(\Lambda) + (\mu^3+1) \sum_{\zeta\in\mathrm{Sin}\Lambda} h(\zeta) + \mu^2 .$$

Proof: for each $v \in \Sigma_f$, we choose $a_v \in k_v = \mathbb{C}_v$ such that $|a_v| \leqq \min_{\zeta\in\mathrm{Sin}\Lambda\setminus\{0\}} |\zeta|_v$. For any matrix U whose entries belong to $K((x))$, we set $U_v(x) := (i_vU)(a_vx)$, where i_v is the embedding $K \hookrightarrow k_v$ (acting on coefficients). Making the change of variable $x \longmapsto a_vx$, we get the p(v)-adic differential operator $\Lambda_v = \frac{1}{a_v} d/dx - \Gamma_v$ or else $L_v = \partial - G_v$ for $G = x\Gamma$, which is better suited to our purpose; then $(G_v)_{[n]} = (G_{[n]})_v$ and the uniform part can be taken as Y_v . We have $R_v(Y) = |a_v|R(Y_v)$. Let us now relate $R_v(\Lambda)$ to $R(L_v)$. To this aim, we bring again the polynomials $q_{a.\cos}$ and $q_{\sin}$ from IV 5.4 into the picture.

We may assume that $\rho(\Lambda) < \infty$ (otherwise the required inequality is empty or trivial). By Katz's theorem, Λ and L_v are in the Fuchsian class (with rational exponents), so that

$$\left.\begin{array}{l} q^b_{a.\cos}(\Lambda)q^{b+m}_{\sin}(\Lambda)\Gamma_m \\[2ex] q^b_{a.\cos}(L_v)q^{b+m}_{\sin}(L_v)\dfrac{x^{-m}G_{v[m]}}{m!} \end{array}\right\} \in M_\mu(K[x]) \quad \text{for a}$$

suitable constant b (we allow finite extension of K). Note that $q^b_{a.\cos}(L_v)q^{b+m}_{\sin}(L_v) = (q^b_{a.\cos}(\Lambda)q^{b+m}_{\sin}(\Lambda))_v$ with the above notation.

Thus $$\left\| \frac{x^{-m}G_{v[m]}}{m!} \right\|_v \leqq \frac{|q^b_{a.\cos}(\Lambda)q^{b+m}_{\sin}(\Lambda)|_{v,0}(1)}{|q^b_{a.\cos}(\Lambda)q^{b+m}_{\sin}(\Lambda)|_{v,0}(|a_v|)} \| a^m_v\Gamma_m\|_v ,$$

which yields the bound

$$(5.1) \quad h_{v,n}(L_v) \leqq \sum_{\mathrm{Sin}\Lambda\setminus\{0\}} \log^+|\zeta|^{-1} + h_{v,n}(\Lambda) + o(n)$$

so that $\log 1/R(L_v) \leq \log 1/R_v(\Lambda) + \sum_{\sin\Lambda\setminus\{0\}} \log^+|\zeta|^{-1}$. Now by reduction to a differential equation (as in the preceding proof) and shearing, we may assume that $G \in M_\mu(K[x]_{(x)})$, and that Λ is normalized. By removing singularities according to § 1, one can change G_v so that the new matrix G_v belongs to $M_\mu(E_{0v})$. We shall distinguish two cases:

1) $R(L_v) > |p|^{1/p}$; then proposition 4.2 applies and yields

$$\log^+ 1/R(Y_v) \leq \frac{\mu^2 \log|p|_v^{-1}}{p(p-1)} + \mu^2(\mu-1)\log 1/R(L_v) .$$

2) $R(L_v) \leq |p|^{1/p}$; lemma 4.1 still applies and yields, after reduction to a differential equation (as in proposition 4.2):

$$\log^+ 1/R(Y_v) \leq \frac{\mu^2 \log|p|_v^{-1}}{p-1} + \mu^2(\mu-1)\log 1/R(L_v) .$$

By hypothesis $|p|^{1/p-1} \geq |p|^{1/p(p-1)} R(L_v)$, hence

$$\log^+ 1/R(Y_v) \leq \frac{\mu^2 \log|p|_v^{-1}}{p(p-1)} + \mu^3 \log 1/R(L_v) .$$

Our three inequalities

$$\begin{cases} \log^+ 1/R_v(Y) = \log 1/R(Y_v) + \log \operatorname{Max}_{\operatorname{Sin}\Lambda\setminus\{0\}} |\zeta|_v^{-1} \\ \log^+ 1/R(Y_v) \leq \dfrac{\mu^2 \log|p|_v^{-1}}{p(p-1)} + \mu^3 \log 1/R(L_v) \\ \log 1/R(L_v) \leq \log 1/R_v(\Lambda) + \sum_{\operatorname{Sin}\Lambda\setminus\{0\}} \log^+|\zeta|^{-1} \end{cases}$$

can be combined and give

$$\rho_f(Y) < \mu^3 \rho(\Lambda) + (1+\mu^3) \sum_{\operatorname{Sin}\Lambda\setminus\{0\}} h_f(\zeta^{-1}) + \mu^2 .$$

We conclude by remarking that $\rho_\infty(Y) = \sum_{v\in\Sigma_\infty} \operatorname{Max}_{\operatorname{Sin}\Lambda\setminus\{0\}} \log^+|\zeta|_v^{-1}$ (see remark 4.4), hence $\rho_\infty(Y) \leq \sum_{\operatorname{Sin}\Lambda\setminus\{0\}} h_\infty(\zeta^{-1})$. □

REMARK. We get a bound for $\rho(Y,Y^{-1})$ this way, if we replace $\operatorname{Sin}\Lambda$ by the set of all finite singularities of Λ .

§ 6. FROM $\rho(Y)$ TO $\sigma(Y)$

6.1 - A "converse" to proposition 4.2

Let k be a ultrametric field as in IV 1, and let G be a matrix such that

i) $G \in M_\mu(k(x) \cap E_0)$,

ii) the eigenvalues α_i of $G(0)$ belong to $\mathbb{Z}_p$,

iii) $L = \partial - G$ is in the Fuchsian class (over $\mathbb{P}^1_k$).

PROPOSITION. Under i) ii) iii), we have the inequality $R(L) \geq R(Y)^{|\mathrm{SinL}|}$, where Y denotes the uniform part in $GL_\mu(k((x)))$ of a solution Yx^C of L .

Proof: recall formula (10) from chapter III, which can be read (for $Y \in GL_\mu(k[[x]])$)

$$(6.1.1)\quad \frac{G_{[m]}}{m!} = \sum_{n\geq 0}\ \sum_{h=0}^{n} Y_h \binom{C+h}{m} (Y^{-1})_{n-h}\, x^n .$$

By shearing we may assume that L is normalized so that such a Y does exist in $GL_\mu(k[[x]])$, and conditions i) ii) iii) are preserved. We may also assume that $C = G(0)$ has Jordan's normal form. By condition iii), we have (with our usual notations) $q^b_{a.\cos}\, q^{b+m}_{\sin}\, x^{-m} \frac{G_{[m]}}{m!} \in M_\mu(k[x])$ for some positive integer b , and the degree of it entries are bounded by $m(|\mathrm{SinL}|-1)+c$ for some other positive integer c . We shall set

$$u_m := q^b_{a.\cos}\, q^{b+m}_{\sin} \Bigl(\prod_{\substack{\zeta\in a\ \mathrm{CosL}\\ \zeta\neq 0}} \zeta^{-b} \Bigr)\Bigl(\prod_{\substack{\zeta\in \mathrm{SinL}\\ \zeta\neq 0}} \zeta^{-m-b} \Bigr)$$

so that a) $u_m \in k[x]$, and the degrees of the entries of $u_m G_{[m]}$ are bounded by $m|\mathrm{SinL}|+c$

b) the first non-zero coefficient of u_m , say u_{m,n_m} , is ± 1 , and $n_m \leq m+b$ (or b if $0 \notin \mathrm{SinL}$).

c) $|u_m| \leq 1$, because $G \in M_\mu(E_0)$ by hypothesis.

We can write (6.1.1) in the following form:

$$(6.1.2)\quad \frac{u_m x^{-m} G_{[m]}}{m!} = \sum_{n=0}^{m(|\mathrm{SinL}|-1)+c}\ \sum_{h+i+j=n+m} Y_h \binom{C+h}{m} (Y^{-1})_i\, u_{m,j}\, x^n .$$

Using b) and c) above, we get

$$\left\| \left(\frac{x^{-m} G_{[m]}}{m!} \right)_n \right\| \leq \frac{\displaystyle \operatorname*{Max}_{h+i \leq m|\sin L|+b+c} \| Y_h \| \, \| (Y^{-1})_i \|}{\displaystyle \operatorname*{Max}_{h \leq m\ |\sin L|+b+c} \left\| \binom{C+h}{m} \right\|} ,$$

for any n, so that

$$R(L) \geq R(Y,Y^{-1})^{|\sin L|} \cdot \lim_{m\to\infty} \left(\operatorname*{Max}_{h\leq m} \left\| \binom{C+h}{m} \right\| \right)^{-\frac{|\sin L|}{m}} .$$

Because $G \in M_\mu(E_0)$, det Y does not vanish inside its convergence disk; indeed det Y satisfies Liouville's equation $\partial(\det Y) = (\mathrm{tr}(G-C))\det Y$. Hence $R(Y,Y^{-1}) = R(Y)$, and it remains to prove that $\lim_{m\to\infty} \left(\operatorname*{Max}_{h\leq m} \left\| \binom{C+h}{m} \right\| \right)^{-1/m} \geq 1$ when $|C| \leq 1$ and the eigenvalues of C belong to $\mathbb{Z}_p$. To this aim, we write $C+h = D+N$, where D denotes the diagonal part of C. Then

$$\binom{C+h}{m} = \sum_{i=0}^{\mu-1} \frac{N^i}{m!} \sum_{j_1<\dots<j_i<m} \prod_{j \neq j_i} (D-j) , \text{ hence}$$

$$\left\| \binom{C+h}{n} \right\| \leq \operatorname*{Max}_{\substack{j_1<\dots<j_l<m \\ l<\mu \\ \alpha \text{ eig. of } C}} \left| \frac{1}{m!} \prod_{j\neq j_i} (\alpha-j) \right| .$$

The result now follows from the following observations:

i) $\alpha, \alpha-1, \dots, \alpha-j, \dots, \alpha-m+1$, for $j \neq j_i$, is a set of $l<\mu$ sequences of "consecutives" elements of $\mathbb{Z}_p$, whose total length is $> m-\mu$,

ii) any sequence of n "consecutives" elements of $\mathbb{Z}_p$, say $\beta, \beta-1, \dots, \beta-n+1$, satisfies $\prod |\beta-j| \leq |p|^{\sum_k ([n/p^k]-1)}$,

iii) $\left| \frac{1}{m!} \right| < |p|^{-\frac{m}{p-1}}$.

6.2 - Statement of the theorem

Let K be a number field. Let $\Lambda = d/dx - \Gamma$ be a differential operator with $\Gamma \in M_\mu(K(x))$, such that a solution at 0 can be expressed by Yc^X, with $Y \in GL_\mu(K((x)))$ and $C \in M_\mu(K)$.

THEOREM. Assume that the eigenvalues of C are rational numbers, with common denominator N, and that Λ is in the Fuchsian class (over $\mathbb{P}^1_{\overline{\mathbb{Q}}}$). Then we have $\sigma(Y) \leq \rho(Y) + N(\mu-1)(\mu^2+1)$. Furthermore, if 0 is a cosingularity, then $\sigma(Y) \leq \rho(Y) + \mu-1$.

The proof is divided into several steps and will occupy the rest of this paragraph.

6.3 - First reductions

We shall prove in fact the following more general result. For each finite place v of K, let k_v denote a complete algebraically closed extension of $\mathbb{C}_v$, and let $Y_v \in GL_\mu(k_v[[x]])$ and $C_v \in M_\mu(k_v)$ such that:

1) $R(Y_v) \geq 1$,
2) $G_v := \partial Y_v \cdot Y_v^{-1} - Y_v C_v Y_v^{-1} \in M_\mu(k_v(x) \cap E_{0,v})$,
3) $L_v := \partial - G_v$ belongs to the Fuchsian class (over $\mathbb{P}^1_{k_v}$),
4) $\| G_v \|_v \leq 1$,
5) C_v is nilpotent.

Under these conditions, we shall prove that

$$(6.3.1)\quad \lim_{l\to\infty} \overline{\lim_{n\to\infty}} \sum_{p(v)\geq l} h_n(Y_v) \begin{cases} \leq (\mu-1)(\mu^2+1) & \text{in the general case,} \\ \leq \mu-1 & \text{if } C_v = 0, \end{cases}$$

where $h_n(Y_v)$ stands for $\frac{1}{n}\log^+ \operatorname{Max}_{m\leq n} \| Y_{v,m} \|_v$. (Note that h_n is the logarithm of a semi-norm on $M_\mu(k_v[[x]])$). Granted (6.3.1), let us now deduce theorem 6.2.

Firstly, we may assume that $\rho(Y) < \infty$, and we remark that:

$$\sigma(Y) = \overline{\lim_{n\to\infty}} \sum_v h_{v,n}(Y) \leq \sum_{\substack{v\in\Sigma_\infty \\ \text{or} \\ p(v)<l}} \overline{\lim_{n\to\infty}} h_{v,n}(Y) + \overline{\lim_{n\to\infty}} \sum_{p(v)\geq l} h_{v,n}(Y)$$

$$(6.3.2)$$

$$= \sum_{\substack{v\in\Sigma_\infty \\ \text{or} \\ p(v)<l}} \log^+ 1/R_v(Y) + \overline{\lim_{n\to\infty}} \sum_{p(v)\geq l} h_{v,n}(Y)$$

for any l, hence it is enough to prove that

(6.3.3) $$\lim_{l\to\infty} \overline{\lim_{n\to\infty}} \sum_{p(v)\geq l} h_{v,n}(Y) \leq \begin{cases} N(\mu-1)(\mu^2+1) & \text{in the general case,} \\ \mu-1 \ , & \text{if 0 is a cosingularity.} \end{cases}$$

Let us reduce the system to a differential equation (after a translation towards an ordinary point, and before going back to 0); thereby we get a differential operator $L = \partial - G$ such that $G \in M_\mu(K[x]_{(x)})$. Since the new matrix Y is obtained from the old one by means of transformation matrices with entries in $K(x)$, $h_{v,n}(Y)$ remains unchanged for all $p(v) >> 0$. Hence in order to prove (6.3.2), we may replace Y by the new one.

We now pass to the variable x^N , so that $NG(x^N)$ replaces G and has integral eigenvalues at 0; we then shear the differential system in order to give it a normalized form, namely with (the new) G nilpotent. By the same argument as above, $h_{v,n}(Y)$ is "essentially" divided by N through these operations, at least for $p(v) >> 0$ (independently of n).

Hence it suffices to prove (6.3.3) with $N = 1$, $G(0)$ nilpotent. Note that $G \in M_\mu(E_{0,v})$ and $\|G\|_v \leq 1$ for $p(v) >> 0$. Now let $a_v \in k_v$, such that $|a_v|_v \leq R_v(Y)$ and $\sum_{v\in\Sigma_f} \log^+|a_v|_v^{-1} < \infty$.

We put $Y_v := Y(a_v x^N)$, $C_v = G(0)$, so that $G_v = G(a_v x)$. All conditions 1)-5) are satisfied, and we have $h_n(Y_v) \geq h_{v,n}(Y) - \log^+|a_v|_v^{-1}$. Hence (6.3.3) for $N = 1$ follows from (6.3.1)

6.4 - <u>Frobenius inverse, twice</u>

According to the proposition 6.1, it follows from conditions 1),2),3),5) above that $R(L_v) = 1$. We also have $\|G_v\| \leq 1$. According to 3.3, the required conditions are now at hand in order to invert the Frobenius functor. We get a matrix $H_v^{inv} \in GL_\mu(E_{0,v})$ (see the remark at the end of 3.3), such that $\|H_v^{inv}\|_v \leq 1$, $H_v^{inv}[G_v] = p\, F_v^\phi$, for some $F_v \in M_\mu(E_0)$; moreover $Y_v = H^{inv}\, Y_{F_v}^\phi$ (since $H^{inv}(0) = I$), so that

(6.4.1) $$h_n(Y_v) \leq \frac{1}{p} h_{[n/p]}(Y_{F_v}) \ .$$

On the other side

$$\| F_v\|_v = |p|_v^{-1} \| \psi_0(H_v^{inv}[G_v])\|_v \leq |p|_v^{-1} \| H_v^{inv}[G_v]\|_v$$

(see the proof of lemma 2.1)

$$\leq |p|_v^{-1} \| H_v^{inv}\|_v \| H_v^{inv-1}\|_v \operatorname{Max}(1, \| G_v\|_v)$$

$$\leq |p|_v^{-1} \text{ in the present case.}$$

If one tries to apply lemma 4.1 to Y_{F_v}, a term $\log\| F_v\|_v$ appears in the estimates, and thus a term $\frac{\log|p|^{-1}}{p}$ appears in the bound for $h_n(Y_v)$. In order to overcome the non-convergence of $\frac{\Sigma \log|p|^{-1}}{p}$, we shall apply the ψ functor again, but to the dual of the ∂-module associated to $\partial - F_v$. In other words, we apply a transformation, say H_v^{inv*}, to $\partial + {}^tF_v$ such that:

$H_v^{inv*} \in GL_\mu(E_{00,v})$, $(H_v^{inv*})^{-1} \in M_\mu(E_{0,v})$,

$H_v^{inv*}[-{}^tF_v] = -p\,{}^tE_v^\phi$ for some $E_v \in M_\mu(E_{00,v})$,

the non-zero singularities of $-{}^tE_v$ are apparent,

$$Y_{F_v} = {}^tY^{-1}_{-{}^tF_v} = {}^t\left(H_v^{inv*}\right)^{-1}\left({}^tY^{-1}_{-{}^tE_v}\right)^\phi = \left({}^tH_v^{inv*}\right)^{-1} Y^\phi_{E_v}.$$

On the other side

$$\| (H_v^{inv*})^{-1}\|_v \leq \operatorname{Max}(\| F_v\|_v, 1)^{\mu-1} \quad \text{by } 3.2.2,$$

$$\leq |p|_v^{1-\mu},$$

and

$$\| E_v\|_v \leq |p|_v^{-1} \| H_v^{inv,*}\|_v \| (H_v^{inv,*})^{-1}\|_v \operatorname{Max}(1, \| F_v\|_v);$$

since $\det(H_v^{inv,*})^{-1}(0) = 1$ and $\det(H_v^{inv,*})^{-1} \in E_{0,v}$. It follows that $\| E_v\|_v \leq |p|_v^{-2-(\mu-1)\mu}$, and:

$$(6.4.2) \qquad h_n(Y_{F_v}) \leq \frac{(\mu-1)}{n} \log|p|^{-1} + \frac{1}{p} h_{[n/p]}(Y_{E_v})$$

for any $n > 0$, and any $v|p$ in Σ_f .
Making use of the matrix $H_v^{cores} \in GL(E_{00,v})$ constructed in remark 1.1, we have the following situation at hand:

$$H_v^{cores}[E_v] \in M_\mu(E_{0,v}), \quad \| H_v^{cores}[E_v] \|_v \leq \| E_v \|_v \leq |p|_v^{-2-(\mu-1)\mu} ,$$

and $\| H_v^{cores} \| \leq 1$, $\| H_v^{cores-1}(0) \| \leq 1$.

The difficulty with the formula
$Y_{E_v} = H_v^{cores} Y_{H_v^{cores}[E_v]} H_v^{cores-1}(0)$ is that $H_v^{cores} \notin M_\mu(E_{0,v})$ in general. However, let $q_v \in k_v[x]$ be a denominator for H_v^{cores} , with $q_v(0) = 1$; then, the previous inequality can be multiplied by q_v , and yields

$$h_n(Y_{E_v}) \leq \frac{\log|q_v|_v}{n} + h_n\left(Y_{H_v^{cores}[E_v]}\right) .$$

In order to evaluate $|q_v|_v$, note that we may choose q_v such that $\det(H_v^{inv,*})^{-1} = q_v e_v$, with e_v invertible in $E_{0,v}$; the point is that local solutions of E_v and F_v are related by $X_{-t_{E_v}} = (H_v^{inv,*})^{-1} X^\phi_{-t_{F_v}}$, or else (more accurately):

$$\det X_{-t_{E_v},a^p} = (\det(H_v^{inv,*})^{-1}\det X_{-t_{F_v},a^\varphi})^{\varphi^{-1}} \circ \theta_a(x-a^p)$$

(with the notation of 3.3.3), and that (for $a \neq 0$) $\det X_{-t_{F_v},a^p}$ has no zero in $D(0,1)$. Since $\det(H_v^{inv,*})^{-1}(0) = 1$, we get $e_v(0) = 1$, hence since e_v is invertible in $E_{0,v}$),

$$|q_v|_v = |\det(H_v^{inv,*})^{-1}||e_v| = |\det(H_v^{inv,*})^{-1}| \geq |p|^{-\mu^2(\mu-1)} ,$$

which gives

$$(6.4.3) \quad h_n(Y_{E_v}) \leq \frac{\mu^2(\mu-1)}{n} \log|p|_v^{-1} + h_n(Y_{H_v^{cores}[E_v]}) .$$

Now lemma 4.1 applies and gives the bound

$$h_n(Y_{H_v^{cores}[E_v]}) \leq \frac{\mu^2\log|p|_v^{-1}}{p-1} + \mu^2\log^+\|H_v^{cores}[E_v]\|_v , \text{ that is to say,}$$

$$(6.4.4) \quad h_n(Y_{H_v^{cores}[E_v]}) \leq (3 + \mu(u-1)\mu^2 \log|p|_v^{-1}.$$

Putting (6.4.1)-(6.4.4) together, we find

$$(6.4.5)\quad h_n(Y_v) \leq \frac{(\mu^2+1)(\mu-1)}{n} \log|p|_v^{-1} + \mu^2(\mu^3+3)\frac{\log|p|_v^{-1}}{p^2} .$$

Now it is clear on formula III(12) that $h_n(Y_v) = 0$ for $p(v)>n$, using properties 2) 4) and 5) in (6.3). on the other side,

$$\sum \frac{\log|p|_v^{-1}}{p^2} < \infty .$$

Thus the required inequality (6.3.1) (in the general case) follows as a straightforward consequence of (6.4.5) and the fact that $\lim_{n\to\infty} \frac{1}{n} \sum_{p(v)\leq n} \log|p|_v^{-1} = 1$.

This completes the proof of theorem 6.2 in the general case.

6.5 - The case of a cosingularity

We now assume that $C_v = 0$. Then, instead of applying the ψ functor, we may use Dwork-Robba theorem as in the proof of theorem IV 5.2. Arguing exactly as in the proof of proposition IV 2.3, we find

$$\left\| \frac{G_{v[n]}}{n!} \right\| \leq \{n,\mu-1\}_v \operatorname*{Max}_{j<\mu} \left(1, \left\| \frac{G_{v[j]}}{j!} \right\| \right)$$

$$\leq \{n,\mu-1\}_v |(\mu-1)!|_v^{-1} .$$

On the other side, $Y_{v,n} = \frac{G_{v[n]}}{n!}(0)$. Hence $h_n(Y_v) \leq \log\{n,\mu-1\}_v$ if $p(v) \geq \mu$. By the computation made in IV 5.2, we get $\lim_{n\to\infty} \frac{1}{n} \sum_{\mu\leq p(v)\leq n} \log\{n,\mu-1\}_v \leq \mu-1$, which yields the required inequality (6.3.1) in this special case, and completes the proof of the theorem.

6.6 - Conclusion

REMARK 1. The inequality (6.4.5) remains true if we replace Y_v by (Y_v, Y_v^{-1}) (using duality). This allows one to replace Y by (Y, Y^{-1}) in the theorem.

REMARK 2. By choosing $Y_v = X_{t_v'}(a_v(x+t_v))$, where $X_{t_v'}$ denotes a complete solution at a the generic point $t_v' \in k_v$ over $\mathbb{C}_v$, and a_v some element of k_v such that $|a_v| \leq R_v(X_{t_v'})$, we get a new proof for $\rho(\Lambda) < \infty \Rightarrow \sigma(\Lambda) < \infty$, but with the less sharp inequality $\sigma(\Lambda) \leq \rho(\Lambda) + (\mu-1)(\mu^2+1)$.

COROLLARY. Let y be some entry in $\bar{\mathbb{Q}}[[x]]$ of "the" uniform part of the solution at 0 of a G-operator. Then y is a G-function.

Proof: it follows from theorem 5 and 6.2, together with Katz's theorem, that $\sigma(y) < \infty$. Thus it is enough to prove that y satisfies some differential equation (linear, homogeneous, with coefficients in $\bar{\mathbb{Q}}(x)$ as usual): we shall find such an equation of order not greater than the square of the order of the G-operator.
Let us write $X = Yx^C$ for a complete solution of the given G-operator (which represents a ∂-module structure M on $\bar{\mathbb{Q}}(x)^\mu$ endowed with its canonical basis). Let $[C]$ stands for the ∂-module structure on $\bar{\mathbb{Q}}(x)^\mu$ represented (in the canonical basis) by the constant matrix C. It is easy to check that the entries of Y are the components (in the canonical basis) of a solution of the ∂-module $\mathrm{Hom}([C],M)$ in $\bar{\mathbb{Q}}((x))$. We conclude by invoking remark 2 in III 3.

EXERCISES. 1) Refining proposition 6.1, give an upper bound for $\rho(L)$ (resp. $\sigma(L)$) in function of $\rho(Y,Y^{-1})$ (resp. $\sigma(Y,Y^{-1})$). In particular, show that if 0 is an ordinary point and X a complete solution with $\rho(X) < \infty$, then Λ is a G-operator.

2) Refining theorem 6.2, give an explicit bound for "the" common denominator of the n first coefficients of a series which is annihilated by a G-operator.

3) Using Dwork-Robba's theorem instead of proposition 6.1, extend the results of this chapter to the case of differential operators whose coefficients are globally bounded functions (not necessarily rational), cf I 4.

APPENDIX: THE GEOMETRIC SITUATION

Recall from Chapter III the definition of a geometric differential equation (over $k = \bar{\mathbb{Q}}$).

THEOREM. *Let $y \in \bar{\mathbb{Q}}[[x]]$ be a solution of a geometric differential equation. Then y is a G-function, and $R_v(y) = 1$ for almost every place v of $\bar{\mathbb{Q}}$.*

Proof: let Λ be such a differential equation (or "operator"), and assume that $R_v(\Lambda) = 1$ for almost every finite place v of a number field K such that the coefficients of Λ belong to

$K(x)$. Then by results of this chapter, y is a G-function; the fact that $R_v(y) = 1$ for almost every v would follow from the quoted result of Christol [13].

Now let us prove that $R_v(\Lambda) = 1$ for almost every v . By results of chapter II, we know that Λ is a product (in the Weyl algebra $A_1(\bar{\mathbb{Q}})$) of factors of Picard-Fuchs equations relative to smooth proper (geometrically connected) varieties $X/_{K(x)}$. Using IV 2.5, we see that it suffices to prove that, for any such $X/_{K(x)}$, $R_v(H^*_{DR}(X)) = 1$ for almost every finite place v of K .

In fact, for every v such that $X_{K_v} \longrightarrow K_v(x)$ lifts to a proper smooth $\mathcal{O}_{K_v}$-morphism $X \to S$ with geometrically connected fibres, one has $R_v(H^*_{DR}(X)) = 1$, because there is an F-crystal structure on $H^*_{DR}(X) \otimes_{K(x)} E_v$ (see e.g. [35][37]). Here S stands for the complement of some points (with distinct residue classes $\neq o$) in Spec $\mathcal{O}_{K_v}[x]$. For an absolutely non-ramified finite place of K , let φ denote the Frobenius endomorphism of K_v , and ϕ the lifting of φ to E_v discussed in this chapter. It can be checked that ϕ extends to a morphism of S^{an} , i.e. respects the completior of $\mathcal{O}_S$ in E_v . The F-crystal structure referred to is a horizontal isomorphism

$$(H^*_{DR}(X) \underset{K(x)}{\otimes} E_v)^{\phi} \simeq H^*_{DR}(X) \underset{K(x)}{\otimes} E_v ,$$

which is furnished by crystalline cohomology (in the ramified case, one has to use the work of Berthelot-Ogus [5]). One concludes using lemma 2.4.

□

This proof also tells us that any geometric differential equation "is" a scalar G-operator (we do not distinguish between the equation $\Lambda? = 0$ and the operator Λ).

The converse statement is a conjecture, known as Bombieri-Dwork's conjecture: every G-operator comes from geometry (in the sense of II).

In fact, the conjecture is slightly stronger and says that if the p-curvatures of Λ are nilpotent for p running over a set of primes of density one; then Λ should be a geometric differential equation.

Chapter VI Global Methods

The method displayed in § 6.1 of the last chapter enables us to pass from invariants of the "uniform part" and its inverse to invariants of the corresponding differential system (ex.3). Now we shall start from one "injective" solution (see below for the meaning of "injective") and shall deduce information about the differential system. To this aim, we follow and simplify an idea of Chudnovsky [18] (and correct a slight mistake in loc. cit. 8.3), involving Hermite-Padé approximants. We refer the reader to [63] for an extensive survey on this theory; however we shall use only very elementary facts from it, so that our exposition remains self-contained.

The need for global considerations in order to transfer information from a single solution y towards Λ should be clear on examples like $y = \Sigma n!x^n$; indeed, $R_p(y) = 1$ for all p , but $R_p(\Lambda) \leqslant |p|^{1/p-1}$ for almost every p and for any non-zero $\Lambda \in \mathbb{Q}[x,d/dx]$ such that $\Lambda y = 0$. In the last paragraph, we summarize in a single statement, the main results obtained in the second part of this book.

§ 1. ITERATING Λ

Let k be a field of characteristic 0. We let $\Lambda = d/dx - \Gamma$, for some matrix $\Gamma \in M_\mu(k(x))$. We define a sequence of matrices $\Gamma_n \in M_\mu(k(x))$ by the following property:

$\frac{1}{n!} d^n/dx^n - \Gamma_n$ is a left multiple of Λ in $M_\mu(k(x))[d/dx]$.

Hence with our usual notations,

$$\Gamma_n = \frac{x^{-n}}{n!} (x\Gamma)_{[n]} .$$

We also have the rule

(1) $\Gamma_{n+1} = \frac{1}{n+1} (d/dx\ \Gamma_n + \Gamma_n \Gamma)$ in $M_\mu(k(x))$.

LEMMA (Chudnovsky). <u>The following identity holds in the ring</u> $M_\mu(k(x))[d/dx]$:

(2) $\Gamma_n = \sum_{m=0}^{n} \binom{n}{m} \frac{(-1)^m}{n!} (d/dx)^{n-m} \cdot \Lambda^m$.

Proof: by induction. This is a tautology for $n=0$. Assume that $(2)_n$ holds, whence after left multiplication by $\frac{1}{n+1}\, d/dx$ in $M_\mu(k(x))[d/dx]$:

$$\frac{1}{n+1}\, d/dx \cdot \Gamma_n = \Gamma_{n+1} + \frac{1}{n+1} \Gamma_n \Lambda$$
$$= \sum_{m=0}^{n} \binom{n}{m} \frac{(-1)^m}{(n+1)!} (d/dx)^{n+1-m} \cdot \Lambda^m .$$

On the other side, right multiplication of $(2)_n$ by $\frac{1}{n+1}\Lambda$ yields

$$\frac{1}{n+1} \Gamma_n \Lambda = \sum_{m=1}^{n+1} \binom{n}{m-1} \frac{(-1)^{m-1}}{n!} (d/dx)^{n+1-m} \cdot \Lambda^m .$$

Using Pascal's identity $\binom{n}{m} + \binom{n}{m-1} = \binom{n+1}{m}$, one finds $(2)_{n+1}$. □

Let $p_0, \ldots, p_{\mu-1} \in k[x]$ be arbitrary polynomials. We define sequences of rational functions ${}_{nj}R = {}_{nj}R(\Lambda)$ by the formula

(3) ${}_{nj}R = \sum_{h=0}^{n} \sum_{l=0}^{\mu-1} \frac{1}{h!} \left(\frac{d^h}{dx^h} p_l\right)_{lj} \Gamma_{n-h}$ for $n \geq 0$, $j = 0, \ldots, \mu-1$.

A straightforward application of Leibnitz's rule and (1) gives

(4) ${}_{n+1,?}R = \frac{1}{n+1}(d/dx\ {}_{n,?}R + {}_{n,?}R\Gamma)$,

where we denote by ${}_{n?}R$ the row $({}_{nl}R)_{l=0,\ldots,\mu-1}$. It is also an easy matter to check that

(5) ${}_{n,?}R \cdot Z = \frac{1}{n!}\, d^n/dx^n \left(\sum_{l=0}^{\mu-1} p_l z_l\right)$ for any

$$Z = {}^t(z_0, \ldots, z_{\mu-1}) \in \ker \Lambda .$$

Indeed, let K be a differential extension of $(K(x), d/dx)$ in which Λ is solvable. To prove (5), it suffices to replace Z by a full set of solutions of Λ , in $GL_\mu(K)$.

Then $\frac{1}{n!}\, d^n/dx^n\left(\sum_{l=0}^{\mu-1} p_l \cdot {}_{l,?}Z\right) = \sum_{l=0}^{\mu-1}\sum_{h=0}^{n}\left(\frac{1}{n!}\, d^h/dx^h\, p_j\right)\cdot$

$$\cdot\left(\frac{1}{(n-h)!}\, d^{n-h}/dx^{n-h}\left({}_{j,?}Z\right)\right)$$

$$= \sum_{l=0}^{\mu-1}\sum_{h=0}^{n}\left(\frac{1}{n!}\, d^h/dx^h\, p_j\right)\cdot {}_{j,?}\Gamma Z$$

$$= {}_{n,?}R\cdot Z \ ,$$

whence the identity (5), since now Z is invertible.

We next replace Λ by the dual differential operator $\Lambda^* = d/dx + {}^t\Gamma$. Identity (4) can be rewritten

(4)* $\quad {}_{?,n+1}R^* = \frac{1}{n+1}\,\Lambda\,({}_{?,n}R^*)$,

where ${}_{?,n}R^*$ stands for the column ${}^t({}_{n,0}R(\Lambda^*),\ldots,{}_{n,\mu-1}R(\Lambda^*))$. Iterating this formula we find:

(5)* $\quad \frac{(m+i)!}{i!}\,{}_{?,m+i}R^* = \Lambda^m({}_{?,i}R^*)$.

Let $R^{<m>}$ denote the matrix whose columns are given by $\binom{m+i}{m}\,{}_{?,m+i}R^*$ for $i = 0,\ldots,\mu-1$. Then, identity (2) inside $M_\mu(k(x))[d/dx]$ specializes to the following identity inside $M_\mu(k(x))$:

(6) $\quad \Gamma_n R^{<0>} = \sum_{m=0}^{n}\sum_{l=0}^{\mu-1}\frac{(-1)^m}{(n-m)!}\, d^{n-n}/dx^{n-m}\, R^{<m>}$.

§ 2. NON-VANISHING OF A CRUCIAL DETERMINANT

2.1 - We now study conditions which may ensure that the matrix $R^{<0>}$ in the left-hand-side of (6) is invertible, i.e. such that $\Delta := \det R^{<0>} \neq 0$.

Let $\Lambda = d/dx - \Gamma$ be a $\mu\times\mu$ differential operator as before. Let $p_0,\ldots,p_{\mu-1}$ be polynomials, not all p_i being 0; let us construct ${}_{jn}R^*$ and $R^{<n>}$ as before.

The following statement is similar (in fact "dual") to the classical non-vanishing theorem of Shidlovski, see e.g. [55] [7]; it generalizes lemma 8.3 of [18], while correcting a slight mistake there: the assertion "$\det(G_0 + G_1 SR^{-1}) \neq 0$" is false ($G_1 = 0$) .

THEOREM. <u>There is a constant</u> $c_0(\Lambda)$ <u>with the following property. If</u> $Y = {}^t(y_0,\dots,y_{\mu-1})$ <u>is a formal solution in</u> $k[[x]]^\mu$ <u>of</u> $\Lambda Y = 0$ <u>with linearly independent entries over</u> $k(x)$, <u>and if</u>

$$\operatorname*{Min}_{ij} \operatorname{ord}_0(p_i y_j - p_j y_i) \geq \operatorname*{Max}_{i} \deg p_i + c_0(\Lambda) ,$$

<u>then we have</u>

$$R^{<0>} \in GL_\mu(k(x)) \text{ , that is } \Delta \neq 0 .$$

Proof: we shall proceed step by step.

2.2 - Let $\nu = \operatorname{rank} R^{<0>}$. Since not all p_i are 0, we have $\nu \geq 1$. Let I be a subset of $\{0,\dots,\mu-1\}$ of cardinality $|I| = \nu$, such that a ν-minor R_I of the form $({}_{i_j,j}R^*)_{j\in I,\ i_1\dots i_j \in\{0,\dots,\mu-1\}}$ (inside $R^{<0>}$) has rank ν .

Let any permutation of the set $\{0,\dots,\mu-1\}$ act on the rows and columns of Γ^* simultaneously; this is compatible with the corresponding permutation of the components of any Z in $\ker \Lambda^*$, hence after (5), with the same permutation of the components of ${}_{?n}R^*$. Thereby we may assume that $(i_1,\dots,i_j) = (0,\dots,\nu-1)$.

2.3 - Let $Z \in GL_\mu(K)$ be a complete solution of $\Lambda^* Z = 0$, in some differential extension K of $(k(x),d/dx)$.

It follows from (5) above that the Wronskian matrix $W = \left(\frac{1}{i!}\frac{d^i}{dx^i}\sum_{h=0}^{\mu-1} p_{h\cdot hj}Z\right)_{i,j}$ constructed on $\left(\sum_{h=0}^{\mu-1} p_{h\cdot h?}Z\right)$ is nothing but ${}^tR^{<0>}Z$. Hence this Wronskian matrix has rank ν .

Let ϕ denote the map $(T_?) \longmapsto \Sigma p_i T_i$; hence the target of ϕ has dimension ν over k .

Therefore by a new choice of Z , one can reach the situtation where $\phi({}_{?,j}Z) = 0$ for $j = 0,\dots,\mu-\nu-1$. We have $\frac{1}{m!} d^m/dx^m \phi({}_{?,j}Z) = \sum_{i=0}^{\mu-1} {}_{im}R^*{}_{ij}Z$, hence the last displayed equation yields the matrix equation

$$ {}^t({}_{ij}Z)_{\substack{i=0,\dots,\mu-1\\ j=0,\dots,\mu-\nu-1}} ({}_{im}R^*)_{\substack{i=0,\dots,\mu-1\\ m=0,\dots,\nu-1}} = 0 .$$

Recall that $R_I = ({}_{im}R^*)_{\substack{i=0,\dots,\nu-1\\ m\in I}}$ has rank ν ; let us set

$R_{I'} = ({}_{im}R^*)_{\substack{i=0,\ldots,\nu-1 \\ m\notin I}}$, and $B = R_{I'} R_I^{-1}$, and rewrite the last system of equations in the form

$$({}_{ji}Z)_{\substack{i\in I \\ j=0,\ldots,\mu-\nu-1}} + ({}_{ji}Z)_{\substack{i\notin I \\ j=0,\ldots,\mu-\nu-1}} B = 0 .$$

It follows that $({}_{ji}Z)_{\substack{i\notin I \\ j=0,\ldots,\mu-\nu-1}}$ has rank $\mu-\nu$, since so does $({}_{ji}Z)_{\substack{i=0,\ldots,\mu-1 \\ j=0,\ldots,\mu-\nu-1}} = ({}_{ji}Z)_{\substack{i\notin I \\ j=0,\ldots,\mu-\nu-1}} (I_{\mu-\nu}, -B)$.

This allows one to conclude that

$$B = -({}_{ji}Z)^{-1}_{\substack{i\notin I \\ j=0,\ldots,\mu-\nu-1}} ({}_{ji}Z)_{\substack{i\in I \\ j=0,\ldots,\mu-\nu-1}} \in M_{\mu-\nu,\nu}(k(x)) ,$$

when $\nu < \mu$.

2.4 - In this situation, a well-known argument of Shidlovski [55] shows that the degree of all numerators and common denominator of the entries of B is bounded by a constant $c_1 = c_1(\Lambda)$ depending only on Λ . For the convenience of the reader, we give the argument.

Let V denote the finite-dimensional k-vector space generated by the monomials of degree $<\mu^2$ in the quantities ${}_{ji}Z$, $i,j = 0,\ldots,\mu-1$.

Let $(z_1,\ldots,z_q)$ be a basis of V such that $z_1,\ldots,z_p$ remain linearly independent over $k(x)$, and form a basis of $V \otimes_k k(x)$. Let $f_{jh} \in k(x)$ be rational functions such that $z_{p+j} = \sum_{h=1}^{p} f_{jh} z_h$, $j = 1,2,\ldots,q-p$. Now suppose that $f \in k(x)$ is a ratio of two elements in V , whence a relation $\sum_{j=1}^{q} \lambda_j z_j = f \sum_{j=1}^{q} \mu_j z_j$, not all μ_j being 0.

We can translate this equation into

$$f = \frac{\lambda_h + \sum_{j=1}^{q-p} \lambda_{p+j} f_{jh}}{\mu_h + \sum_{j=1}^{q-p} \mu_{p+j} f_{jh}} \quad .$$

Since the rational functions f_{jh} depend only on V which is determined by Λ but is independent of the polynomials $p_0,\dots,p_{\mu-1}$, we are done.

2.5 - Now let $Y = {}^t(y_0,\dots,y_{\mu-1})$ be the particular solution of $\Lambda Y = 0$ being considered in the theorem and let us introduce the $\nu\times\mu$-matrix (in case $\nu<\mu$):

$$\left[\begin{array}{cccc|c} y_{\mu-1} & & & & -y_0 \\ y_1 & -y_0 & & & \\ \vdots & & \ddots & & \\ y_{\nu-1} & & & -y_0 & \end{array}\right]$$

$$\underbrace{\qquad\qquad\qquad\qquad}_{S'} \qquad \underbrace{\qquad\quad}_{S''}$$

Hence the matrix $T := S({}_{i,m_j}R^*)_{\substack{i=0,\dots,\mu-1 \\ m_j \in I}}$ has entries:

$$ {}_{ij}T = \begin{cases} {}_{0,m_j}R^* y_i - {}_{i,m_j}R^* y_0 & \text{for } i>0 \\ {}_{0,m_j}R^* y_{\mu-1} - {}_{\mu-1,m_j}R^* y_0 & \text{for } i=0 \ . \end{cases}$$

We have $TR_J^{-1} = S' + S''B$, and this matrix looks like

$$\left[\begin{array}{cccc} y_{\mu-1}-b_0y_0 & -b_1y_0 & \cdots\cdots & -b_{\mu-1}b_0 \\ y_1 & -y_0 & & \\ \vdots & & \ddots & \\ y_{\nu-1} & & & -y_0 \end{array}\right] .$$

Therefore

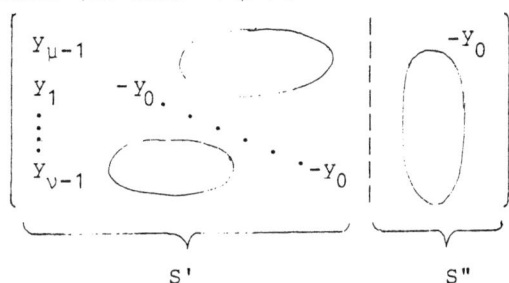

$$\det TR_J^{-1} = \begin{vmatrix} y_{\mu-1} - \sum_{i=0}^{\nu-1} b_i y_i & & & \\ y_1 & -y_0 & & \\ \vdots & & \ddots & \\ y_{\nu-1} & & & -y_0 \end{vmatrix}$$ does not

vanish, since the components y_i of Y are linearly independent over $k(x)$ by assumption.

2.6 - Using formula 4* , we can compute

$$\frac{1}{m+1} d/dx({}_{i,m}R^*y_j - {}_{j,m}R^*y_i)$$

(7)
$$= {}_{i,m+1}R^*y_j - {}_{j,m+1}R^*y_i + \sum_{h=0}^{\mu-1} {}_{ih}\Gamma({}_{h,m}R^*y_j - {}_{j,m}R^*y_h) .$$

Letting $\delta = \mathrm{Max}(1,-\mathrm{ord}_0\Gamma)$, we conclude that $\mathrm{ord}_0\ {}_{ij}T \geq \mathrm{Min}_i(\mathrm{ord}_0({}_{i0}T)) - m_j\delta$, hence

$$\mathrm{ord}_0 \det T \geq \nu \,\mathrm{Min}_i(\mathrm{ord}_0(p_iy_j - p_jy_i)) - \frac{\nu(\nu-1)\delta}{2} .$$

On the other hand, it follows clearly from (4)* (by induction), that

$$(8) \begin{cases} \deg \mathrm{den}\ {}_{im}R^* \leq m.\deg \mathrm{den}\ \Gamma \\ \deg \mathrm{num}\ {}_{im}R^* \leq \max_j \deg p_j + m.(\deg \mathrm{den}\ \Gamma-1) \end{cases}$$

for $m \geq 0$ and $j = 0,\ldots,\mu-1$. Thereby the degrees of all rational function entries of R_J^{-1} are bounded by $\frac{\nu(\nu+1)}{2}c_2(\Lambda) + \nu \,\mathrm{Max}_i \deg p_i$. As for B , we have seen in 2.4 that the degrees of its entries are bounded by $c_1(\Lambda)$. Since the y_i's are components of a solution of Λ , $\mathrm{ord}_0\ y_i$ can also be bounded by a constant $c_3(\Lambda)$ depending only on Λ . We thus find the following double inequality:

$$\mathrm{ord}_0 \det TR_J^{-1} = \mathrm{ord}_0 \det(S'+S''B) \leq \nu(c_1(\Lambda) + c_3(\Lambda))$$

and

$$\mathrm{ord}_0 \det TR_J^{-1} \geq \nu(\underset{ij}{\mathrm{Min}}\ \mathrm{ord}_0(p_iy_j - p_jy_i) - \underset{i}{\mathrm{Max}}\deg p_i) - \frac{\nu(\nu-1)}{2}\delta - \frac{\nu(\nu+1)}{2}c_2(\Lambda) \quad .$$

Now let $c_0(\Lambda)$, the constant in the theorem, be

$$c_0(\Lambda) = c_1(\Lambda) + c_3(\Lambda) + \frac{\mu+1}{2}(\delta+c_2(\Lambda)) \quad .$$

Then $\underset{ij}{\mathrm{Min}}\ \mathrm{ord}_0(p_iy_j - p_jy_i) - \underset{i}{\mathrm{Max}}\deg p_i \geq c_0(\Lambda)$ yields a contradiction, in the case $(\nu<\mu)$ under consideration; hence this inequality forces $\nu=\mu$.

□

This allows one to invert $R^{<0>}$ in formula (6).

§ 3. HERMITE-PADÉ APPROXIMANTS

Looking for polynomials p_i satisfying the assumption of the theorem is the very classical problem of (dual) Hermite-Padé approximants.

Here is a simple construction.

Let $q \in k[x]$ be a non-zero polynomial of degree $\leq N$, satisfying the following condition:

(9) $\mathrm{ord}_0(qY)_{>N} \geq N(1+\frac{1-\tau}{\mu}) + 1$ for some $\tau,\ 0<\tau<1$.

Here we have set $(qY)_{>N} = qY - (qY)_{\leq N} = \sum_{n>N}\sum_{i=0}^{n} q_i x^n Y_{n-i}$.

Condition (9) means that qY should be lacunary between order N and order $N(1+\varepsilon)$, for small ε .

More precisely, (9) is a linear system of $\mu[\frac{N(1-\tau)}{\mu}]$ equations in the $N+1$ unknown quantities $q_0,\dots,q_N$ (the coefficients of q); there always exists a non-zero solution q , since

$$N+1 > \mu[\frac{N(1-\tau)}{\mu}] \quad .$$

In order to satisfy the assumptions of the theorem, it suffices to choose N large enough, and to set

(10) $\quad p_i := (q\ y_i)_{\leq N}$.

Note that, given q , this is the unique choice for which

$\mathrm{ord}_0(q\,y_i - p_i) > N$. Then $\mathrm{ord}_0(p_i y_j - p_j y_i) \geq N(1+\frac{1-\tau}{\mu})+1$.

Let us denote by u a common denominator in $k[x]$ of the entries of Γ , and let $s := \mathrm{Max}(\deg u, 1+\deg({}_{ij}\Gamma)u)$. As previously, we assume that Y is a solution of $d/dx\,Y = \Gamma Y$.

LEMMA. Under (10), we have the relation

$$(11)_m \quad \left(\frac{u^m}{m!}(d^m/dx^m q)\cdot y_i\right)_{\leq N+m(s-1)} = {}_{im}R^*\,u^m \quad \underline{\text{for}} \quad i = 0,\ldots,\mu-1$$

and $m \leq \frac{N(1-\tau)}{\mu s}$.

Proof: of course, the case $m=0$ is a tautology. We first notice that $\deg_{im}R^*\,u^m \leq N+m(s-1)$. By induction, and using the last remark, it suffices to show that $\mathrm{ord}_0({}_{im}L) \geq N+m(s-1)+1$ for ${}_{im}L := \frac{u^m}{m!}\,d^m/dx^m q\cdot y_i - {}_{im}R^*\,u^m$, and for any i, m as in the statement. We can readily compute

$$\frac{u}{m+1}\,d/dx\;{}_{im}L = \frac{u^{m+1}}{(m+1)!}\,d^{m+1}/dx^{m+1}q\cdot y_i + \frac{mu^m}{(m+1)!}\,(d/dx\,u)\cdot d^m/dx^m q\cdot y_i$$

$$+ \frac{u^{m+1}}{(n+1)!}\sum_{j=0}^{\mu-1}{}_{ij}\Gamma\;d^m q/dx^m\cdot y_j - m(d/dx\,u)\cdot\frac{u^m}{(m+1)!}\,{}_{im}R^* - \frac{u^{m+1}}{(m+1)!}\sum_{j=0}^{\mu-1}{}_{ij}\Gamma\,{}_{jm}R^*$$

$$- \frac{u^{m+1}}{(m+1)!}\,{}_{i,m+1}R^*$$

$$= {}_{i,m+1}L + \frac{m(d/dx\,u)}{m+1}\,{}_{im}L + \frac{u}{m+1}\sum_{j=0}^{\mu-1}{}_{ij}\Gamma\cdot{}_{jm}L\ , \text{ which gives}$$

$$\min_i \mathrm{ord}_0({}_{i,m+1}L) \geq \min_i \mathrm{ord}_0({}_{i,m}L) - 1$$

$$\geq N(1+\frac{1-\tau}{\mu}) - m \quad \text{by recursion.}$$

We conclude by noting that

$N+m(s-1)+1 \leq N(1+\frac{1-\tau}{\mu})-m+1$ for m as in the lemma.

□

§ 4. FROM $\sigma(y)$ TO $\sigma(\Lambda)$

Now $k = K$ is a number field. By Siegel's lemma (see I 1.1), we can find a non-zero polynomial $q = \sum_{n=0}^{N} q_n x^n \in \mathcal{O}_K[x]$ which satisfies condition (9), with

$$h(q_0,\ldots,q_n) \le \frac{\mu[\frac{N(1-\tau)}{\mu}]}{N+1-\mu[\frac{N(1-\tau)}{\mu}]}[h(y_{0,0},y_{0,1},\ldots,y_{\mu-1,[N(1-\frac{1-\tau}{\mu})]})$$

$$+ \log N + c(K)] \ ,$$

that is to say, making use of the convention I 1.2:

$$(12) \quad (1+\deg q)h(q) \le \frac{1-\tau}{\tau}[N(1+\frac{1-\tau}{\mu})h(Y_{\le N(1-\frac{1-\tau}{\mu})}) + \log N + c(K)] \ .$$

Without loss of generality, one may also assume that the polynomial u lies in $\mathcal{O}_K[x]$. To go further, we have to fix $n \in \mathbb{N}$, and to assume that

$$(13) \quad N \ge \frac{\mathrm{Max}(c_0(\Lambda),\mu s(n+\mu-1))}{1-\tau}, \quad c_0(\Lambda) \text{ being as in theorem 2.}$$

Now let v be a finite place of K . Under condition (13), the assumption of theorem 2 is fulfilled, hence $R^{<0>}$ can be inverted in formula (6); this gives the following estimate:

$$\|\Gamma_n\|_v \le \mathrm{Max}_{h\le n} \|\frac{1}{h!}(\frac{d^h}{dx^h}R^{<n-h>}) \,\mathrm{Adj}\, R^{<0>}\|_v \,|\det R^{<0>}|_v^{-1}$$

$$\le \mathrm{Max}_{h\le n} \|R^{<h>}\,\mathrm{Adj}\, R^{<0>}\|_v\,|\Delta|_v^{-1} \quad \text{according to IV 1.5.1,}$$

henceforth the following bound

$$\|\Gamma_n\|_v \le (\mathrm{Max}_{h\le n}\|R^{<h>}\|_v)^{\mu}|\Delta^{-1}|_v$$

$$\le (\mathrm{Max}_{\substack{h\le n+\mu-1\\ i\le \mu-1}} |{}_{ih}R^*|_v)^{\mu}|\Delta^{-1}|_v \ .$$

We use lemma 3 in order to estimate $|{}_{ih}R^*|_v$; indeed condition (13) makes $(11)_n$ available. We obtain

$$(14) \quad {}_{ih}R^* = u^{-h} \sum_{h+l+m\le N+h(s-1)} (u^h)_k \binom{h+l}{l} q_{h+l}\, y_{i,m}\, x^{k+l+m} \ ,$$

whence

$$|{}_{ih}R^*|_v \le |q|_v \mathrm{Max}_{m\le N+h(s-1)} |y_{i,m}|_v$$

$$\le \mathrm{Max}_{m\le N+h(s-1)} |y_{i,m}|_v \ , \text{ since } q \in \mathcal{O}_K[x] \ .$$

Therefore we get

(15) $\|\Gamma_n\|_v \leq (\underset{m\leq N+(n+\mu-1)(s-1)}{\mathrm{Max}} |y_{i,m}|_v^{\mu}) \cdot |(u^{\frac{\mu(\mu-1)}{2}}\Delta)^{-1}|_v$

because $u \in \mathcal{O}_K[x]$. Now let ζ be any root of unity satisfying $(u^{\frac{\mu(\mu-1)}{2}}\Delta)(\zeta) \neq 0$. We have

$$\sum_v h_{v,n}(\Lambda) = \sum_{v\in\Sigma_f} \mathrm{Max}(1/n \underset{m\leq n}{\mathrm{Max}} \log\|\Gamma_m\|_v, 0)$$

$$\leq \sum_{v\in\Sigma_f} \underset{m\leq n}{\mathrm{Max}} \frac{1}{n}\log\|\Gamma_m\|_v \quad \text{since } \Gamma_0 = I$$

$$\leq \mu\sigma_f(Y)(\frac{N}{n}+1+o(1)) + \frac{1}{n}\sum_{v\in\Sigma_f}\log|(u^{\frac{\mu(\mu-1)}{2}}\Delta)(\zeta)|_v^{-1} .$$

But the "product formula" in $K\setminus\{0\}$ yields

$$\sum_{v\in\Sigma_f}\log|(u^{\mu(\mu-1)/2}\Delta)(\zeta)|_v^{-1} = \sum_{v\in\Sigma_\infty}\log|u^{\mu(\mu-1)/2}\Delta)(\zeta)|_v .$$

Using formula (14), one finds

$$\frac{1}{n}h_\infty((u^{\mu(\mu-1)/2}\Delta)(\zeta)) \leq \mu(\frac{1}{n}(1+\deg q)h(q) + \frac{N}{n}\sigma_\infty(y)) + o(1)$$

since $h(\zeta) = h(1) = 0$. Putting everything together, we arrive at the upper bound

(15) $\sum_v h_{v,n}(\Lambda) \leq \mu[(1+\frac{N}{n})\sigma(Y) + \frac{1}{n}(1+\deg q)h(q)] + o(1)$.

Now, we choose N to be the <u>minimal</u> integer satisfying condition (13). Hence $\frac{N}{n} \sim \frac{\mu s}{1-\tau}$, when $n \to \infty$. Making use of the estimate (12), we can translate (15) into

$$\sigma(\Lambda) := \overline{\lim_{n\to\infty}}\sum_v h_{v,n}(\Lambda)$$

$$\leq \mu\left[1+\mu s\left(\frac{1}{\tau} + \frac{1-\tau}{\mu\tau} + \frac{1}{1-\tau}\right)\right]\sigma(Y) ,$$

and letting $\tau = 1/2$, we arrive at

(16) $\sigma(\Lambda) \leq 5\mu^2 s\,\sigma(Y)$, if $\mu > 1$ $\quad$ ($\leq (1+5s)\sigma(Y)$ if $\mu = 1$).

The qualitative part of the following conclusion follows readily from (16) and was given by Chudnovsky [18] (for $K = \mathbb{Q}$).

THEOREM. <u>Let</u> $\Lambda = d/dx - \Gamma$ <u>be a differential operator with</u> $\Gamma \in M_\mu(K(x))$. <u>Assume that a solution</u> Y <u>of</u> $\Lambda Y = 0$ <u>in</u> $K[[x]]^\mu$ <u>has linearly independent components over</u> $K(x)$. <u>Then</u>

$\sigma(\Lambda) \leq 6\mu^2 s\, \sigma(Y)$. *In particular* Λ *is a G-operator if the components of* Y *are G-functions.*

Recall that s denotes $\mathrm{Max}(\deg u; 1+\deg({}_{ij}\Gamma)u)$ for a common denominator u of the entries of Γ .

□

For want of a better place, we now give some *algebraic structure* properties of the set of all G-functions. We make our exposition depend on the previous theorem, but this could be avoided using direct recurrence arguments à la Lipshitz.

COROLLARY. *The set of all G-functions is a* $\bar{\mathbb{Q}}$*-sub-vector space of* $\bar{\mathbb{Q}}[[x]]$, *stable under the Cauchy and Hadamard products.*

This statement should be compared to theorem II 3.

Proof: we first notice that the condition $\sigma < \infty$ is stable under $+, \times$ and $*$ according to I 1.
The stability of the condition " y is killed by some non-zero element of $\bar{\mathbb{Q}}[x,d/dx]$ " under $+$ and $\times$ follows formally from Ore's localizability condition in the Weyl algebra $A_1(\bar{\mathbb{Q}}) = \bar{\mathbb{Q}}[x,d/dx]$, as explained in the proof of II 3. It remains to prove that this condition is also stable under $*$, under the *additional* condition $\sigma < \infty$. But according to the proof of the previous theorem, $\sigma(y) < \infty$ implies that any non-zero element Λ of $A_1(\bar{\mathbb{Q}})$ of *minimal* order with respect to the property $\Lambda y = 0$ has to be in the Fuchsian class. The required stability then follows from corollary 2 in I 3.3.

□

SCHOLIE. *The units in the algebra of all G-functions* (under the Cauchy product) *are the elements* $y \in \bar{\mathbb{Q}}[[x]]$ *which are algebraic over* $\mathbb{Q}(x)$ *and satisfy* $y(0) \neq 0$.

Sketch of proof: we adapt an argument of G.Christol [16], who also deals with diagonals. If y and $1/y$ are solutions of linear differential equations, y'/y must be an algebraic function, say g , according to a theorem of W.A. Harris and Y. Sibuya [" the reciprocals of solutions of linear ordinary differential equations", Adv. in Math. 58 (1985)], 119-132.

Let us denote by M the ∂-module over $K(x)$ which arises from the differential equation $y' = gy$; here K stands for any number field which contains the coefficients of y . We have $\dim \mathcal{M} = [K(z,g):K(x)]$.

It follows from the previous theorem that a quotient of $\mathcal{M}$, say $\mathcal{N}$ (for which y becomes an injective solution, cf. next paragraph) satisfies $\rho(\mathcal{N}) < \infty$, therefore this quotient must have nilpotent v-curvatures for every finite place v whose residue characteristic of theorem IV 5.3 and ex. IV 4). On using some elementary differential Galois theory (see e.g. [2]) and the fact that $\mathcal{N}$ actually "comes from" a module of rank one by a restriction of scalars, one may derive that these v-curvatures vanish, in point of fact. A fundamental result of Chudnovsky [19], cf. VIII ex.5, based upon methods to be described in ch. VIII, implies that y is algebraic. The converse statement is immediate. □

The determination of the units under the Hadamard product has not yet been studied; L_1/x is an example (notation of I 4).

REMARK. It follows easily from I 1.4 lemma 2, that the set of G-functions is stable under taking primitive or derivative.

OPEN QUESTIONS. 1) Under the assumptions of the theorem, does the following implication hold true: $\rho(Y) < \infty \Rightarrow \Lambda$ is a G-operator?

2) For any G-function y , does the inequality $\rho(y) \leq \sigma(y)$ hold?

It would be enough to show that for every place v , $\lim_{n\to\infty} h_{v,n}$ does exist (by Fatou's lemma).

D. Bertrand suggests to relate this question to Poincaré's theorem about recurrent series.

§ 5. MAIN THEOREM

We summarize here the main qualitative results of the second part of this book. By injective solution, we mean an injective $A_1(\bar{\mathbb{Q}})$-morphism $\mathcal{M} \hookrightarrow \bar{\mathbb{Q}}((x))$.

MAIN THEOREM. Let M be a ∂-module over $\bar{\mathbb{Q}}(x)$, and let θ be an injective solution of M into $\bar{\mathbb{Q}}((x))$. The following assertions are equivalent

i) for some cyclic element m of M, $\sigma(\Theta(m)) < \infty$

ii) for every $m \in M$, $\sigma(\Theta(m)) < \infty$

iii) $\sigma(M) < \infty$

iv) $\sigma(M^*) < \infty$

v) $\rho(M)$ ($= \rho(M^*)$) $< \infty$

and they imply

vi) for every $m \in M$, $\rho(\theta(m)) < \infty$.

REMARK. Recall that for any $m \in M$, $\theta(m)$ lies in $K((x))$ for some number field K;this gives a sense to $\rho(\theta(m))$. However these invariants do not depend on the choice of K. I don't know whether vi)$\Rightarrow$i).

Proof of the theorem: i) $\Rightarrow$ ii) is a formal consequence of I.1, lemma 2. Indeed let m_{cyc} be a cyclic element and let m be any element of M ; we can write $m = \sum_{i=0}^{\mu-1} f_i \partial^i m_{cyc}$, for suitable rational funtions f_i. Then $\theta(m) = \sum_{i=0}^{\mu-1} f_i \partial^i \theta(m_{cyc})$ because θ is a morphism of $A_1(\bar{\mathbb{Q}})$-modules, and this gives

$$\begin{aligned}\sigma(\theta(m)) &\leq 3/2\ \sigma(f_0,\dots,f_{\mu-1},\theta(m_{cyc}))\\ &\leq 3/2\Big(\sigma(\theta(m_{cyc})) + \sum_{i=0}^{\mu-1}\sigma(f_i)\Big)\end{aligned}$$

hence if $\sigma(\theta(m_{cyc}))$ is finite, so is $\sigma(\theta(m))$, because σ takes finite values on $\bar{\mathbb{Q}}(x)$, see I 4.

ii) $\Rightarrow$ iii): choose a basis $\{m_i\}$ of the $\bar{\mathbb{Q}}(x)$-vector space underlying M such that $\theta(m_i) \in \bar{\mathbb{Q}}[[x]]$ for every $i=0,\dots,\mu-1$. Since θ is injective, the series $\theta(m_i)$ are linearly independent over $\bar{\mathbb{Q}}(x)$, and formula (16) above gives the required implication.

iii) $\Leftrightarrow$ iv): this is proposition 5.5 of chapter IV.

iii) $\Rightarrow$ v): this is part of theorem IV 5.2.

v) $\Rightarrow$ i): this is a slight extension (for $y \in \bar{\mathbb{Q}}((x))$ instead of $\bar{\mathbb{Q}}[[x]]$) of corollary V 6.6, with the same proof.

v) ⇒ vi): this is included in theorem V 5, once we have remarked that a solution ${}^t(y_0=\theta(m_0),\dots,y_{\mu-1}=\theta(m_{\mu-1}))$ is always the first column of a suitably chosen "uniform part" of the complete solution at 0 of the differential system associated to M equipped with basis $\{m_i\}$. Indeed by Katz's theorem, a complete solution may be written in the form $Y\,x^C$, and our particular solution is expressed by $Y\,x^C E_{11}$, with

$$E_{11} = \begin{pmatrix} 1 & 0 \\ 0 & 0 \end{pmatrix}$$

, i.e. by the first column of $Y\,x^C$ which can also be rewritten $x^\alpha Y\,E_{11}$ for some "exponent" α . It is clear that we may assume that $\alpha=0$ without loss of generality (there is much freedom in the choice of Y).

□

REMARKS. 1) If M is an irreducible ∂-module over $\bar{\mathbb{Q}}(x)$, then the non-zero solution of M is of course injective.

2) It is plain that the theorem fails if θ is not injective, even if it is non-zero. E.g., let

$M = (\bar{\mathbb{Q}}(x), \partial 1=0) \oplus (\bar{\mathbb{Q}}(x), \partial 1=x)$ and θ = first projection.

COROLLARY. <u>Every G-function</u> y <u>satisfies</u> $\rho(y) < \infty$.

Proof: this is a plain consequence of i) ⇒ vi) in the theorem.

□

REMARK. Recall from I 5 that $\sigma(y) < \infty$ does not imply $\rho(y) < \infty$ in general.

EXERCISES. 1) Relate the exponents of the (generalized) hypergeometric differential equation in I 4.4 to the parameters of the hypergeometric function solution y . Show that $\sigma(y) < \infty$ implies that these parameters are rational numbers.

2) Use the zero estimate given in III Appendix, to compute explicitely a constant $c_0(\Lambda)$ as in theorem 2 in case $\sigma(Y) < \infty$.

3) For any G-function y , give a bound for $\sigma(y)$ in terms of $\rho(y)$ which looks like the one in theorem V 6.2. (Hint: use ibid. 6.3.2).

Part Three

Diophantine Questions

Chapter VII Independence of Values of G-Functions

§ 1. INTRODUCTION

In his paper of 1929 [56], C.L. Siegel, after defining G-functions and giving some examples, announced some results which one could obtain by the techniques he found (and described in the same paper) for studying the diophantine approximation properties of values of what he called E-functions. However no proof had appeared, and the first attempt in the direction of Siegel's statements was in M.S. Numagomedov's work, more than fourty years later. The successive work of A.I. Galočkin [30], Y. Flicker, E. Bombieri [7] and D.V. & G.V. Chudnovsky [18], finally completed the proof of a G-function theorem that Siegel could have envisioned; roughly speaking, this is a quantitative result on the non-existence of too many algebraic relations among the values at some algebraic point ξ of certain G-functions, when ξ is "arithmetically" small enough (depending on the degree of the relations). Therefore, the diophantine theory of values of G-functions belongs to irrationality rather than transcendence theory. Nevertheless, its typical feature (and strength), discovered by Bombieri, is the possibility of a local-to-global setting. This leads to a kind of Hasse's principle for values of G-functions as follows: calling "global" a dependence relation which is true in a p-adic field whenever it makes sense in that field, then every global relation at ξ comes from a functional relation between G-functions $y_0,\dots,y_{\mu-1}$, except for a set of ξ of bounded height.

All of the works mentioned above use some invariants of Siegel's method. Here we shall simplify and sharpen substantially all those results by using instead Gel'fond's method, following an idea of P. Debes [20]. We shall also make the constants explicit.

Nevertheless, the reader will profitably compare the two methods (see ex.3 below). Indeed, following a way traced by S. Lang in the E-function case [42], Siegel's method also gives rise to lower bounds for polynomials in special values of G-functions, a subject that we have not at all touched upon, because of the present lack of geometric applications - we refer to the works of D.V. and G.V. Chudnovsky [18], K. Väänänen [60] and Xu G. for this matter.(1)

In this chapter, we shall also illustrate the powerful local-to-global method, but only in the special case of algebraic functions, by giving two applications: firstly, an effective version of Hilbert's irreducibility theorem (after V. Sprindzuk), and then a study of certain rational points over superelliptic curves.

In this latter application of the Hasse principle for values of G-functions, we obtain effective bounds of a new kind; the method itself can be viewed somehow as an "adelized" version of the old method of Runge - the first general statement about two-variables diophantine equations - which we recall briefly in the appendix.

---

(1) This theory has a long-standing bad reputation of non effectivity, which comes from the constant $c_0(\Lambda)$ in Shidlovski's theorem (VI 2.1, or the more usual "dual" statement [55] [7]). However, this can be overcome by using instead III appendix, taking into account the fact that a differential system satisfied by a column of linearly independent G-functions is Fuchsian (VI 4). Also the needed explicit estimates for the coefficients of the G-functions may be derived from the asymptotic estimate (for σ, see V ex.2).

§ 2. APPROXIMATING FORMS

2.1 - Let K be a number field. Let $\Gamma \in M_\mu(K(x))$ and let us consider the differential operator $\Lambda = d/dx - \Gamma$; we set $s := |\mathrm{Sin}\Lambda|$. Let $Y = {}^t(y_0,\dots,y_{\mu-1})$ be a solution of $\Lambda Y = 0$ in $K[[x]]^\mu$, with linearly independent entries (over $K(x)$). We assume that $\sigma(Y) < \infty$, or equivalently, that the y_i's are G-functions.

At least, let $\xi \neq 0$ be an algebraic number in K , which is an ordinary point or an apparent singularity of Λ .

2.2 - Let $Z \in M_{\mu\nu}(K[[x-\xi]])$, $0 < \nu \leqq \mu$, be a solution of $\Lambda Z = 0$ (since ξ is at worst an apparent singularity, one could even take $\nu = \mu$ and Z invertible, but in the application in view, Z will have rank $\nu < \mu$). We shall construct a row of polynomials $P = (p_0,\dots,p_{\mu-1})$ such that $P.Z$ has a high order at ξ .

LEMMA. Let α be a positive integer, and let τ be a positive real number. Then there exists $P = (p_0,\dots,p_{\mu-1}) \in K[x]^\mu$, with the following properties:

i) $\deg P \leqq \frac{\nu\alpha}{\mu\tau}(1+\tau)$, p_i not all 0,

ii) $$(1+\deg P)\frac{h(P)}{\alpha} \leqq \left\{\frac{\nu}{\mu}(1+\tau)+\tau(2s-1)\right\}h(\xi)$$
$$+\left\{s\tau+\frac{\nu}{\mu}(1+\tau)\right\}\log 2 + 2\tau\sum_{\mathrm{Sin}\Lambda} h(\zeta) + \tau\sigma(\Lambda) + o(1)$$

iii) $\mathrm{ord}_\xi PZ \geqq \alpha$,

where $o(1)$ is relative to $\alpha \to \infty$.

REMARK: $\sigma(\Lambda) < \infty$ because $\sigma(Y) < \infty$ and the y_i's are linearly independent, according to the results of the last chapter.

Proof: condition iii) is the following linear system of $\nu\alpha$ equations in the μN unknowns p_{in}, $i=0,\dots,\mu-1$; $n=0,\dots,N-1$, with $N := \left[\frac{\nu\alpha}{\mu\tau}(1+\tau)\right] + 1$:

$$\sum_{n=0}^{N-1}\sum_{i=0}^{\mu-1}\sum_{l=0}^{\mathrm{Min}(m,n)}\binom{n}{l}\xi^{n-l}p_{im}\cdot{}_{ij}Z_{m-l} = 0,\ j=0,\dots,\nu-1;\ m=0,\dots,\alpha-1 .$$

Hence by Siegel's lemma, there exists a non-zero solution with

$$h((p_{in})_{i,n}) \leq \frac{\nu\alpha}{\mu N-\nu\alpha}\{(N-1)(h(\xi)+\log 2) + \alpha\, h(Z_{\leq\alpha}) + \log \mu N^2 + c(K)\}$$

$$\leq \tau\left\{\frac{\nu\alpha}{\mu\tau}(1+\tau)(h(\xi)+\log 2) + \alpha\, \sigma(Z)\right\} + o(\alpha) ,$$

and we conclude using proposition IV 5.4 (recall our convention for the height of polynomials).

□

2.3 - Let us now consider the series $r := \sum_{i=0}^{\mu-1} p_i y_i$. Since $\sigma(\Lambda) < \infty$, it follows from results of chapter IV that Λ is Fuchsian. Hence the zero estimate in chapter III (appendix) is available in our present case, and yields

$$\beta := \mathrm{ord}_0 r \leq \mu(N-1)+o(1) \leq \frac{\nu\alpha}{\tau}(1+\tau) + o(1) .$$

§ 3. A METHOD OF GEL'FOND

3.1 - Let us consider the first non-zero coefficient of r :

$$\eta := \frac{1}{\beta!}\left.\frac{d^\beta r}{dx^\beta}\right|_0 = \sum_{i=0}^{\mu-1}\sum_{n=0}^{\beta} p_{i,n} y_{i,\beta-n} \in K\setminus\{0\} .$$

For any place v of K , we have,

$$(3.1.1)\quad \log|\eta|_v \leq \operatorname*{Max}_{i,n} \log^+|p_{i,n}|_v + \beta\, h_{v,\beta}(Y) ,$$

with an additional term $(d_v/d)\log \beta\mu$ if $v \in \Sigma_\infty$.

3.2 - We now assume that there exist $\mu-\nu$ linearly independent relations over K

$$(3.2.1)\quad \sum_{i=0}^{\mu-1} a_{ij}\, y_i(\xi) = 0,\ j=0,\dots,\mu-\nu-1 .$$

Such relations may be understood in several ways. Indeed the functions y_i may be considered as convergent Taylor series $i_v(y_i)$ in the v-adic disk $D_v(0,R_v(Y))$. Thus of $|\xi|_v < R_v(Y)$, it makes sense to consider the above relations as holding in K_v .

Now we assume that (3.2.1) is satisfied <u>v-adically for every</u> v <u>in some fixed finite set of places</u> V , <u>such that</u>

$$|\xi|_v < \mathrm{Min}(1,R_v(Y))\,\mathrm{Min}(1,|2|_v^{-1}), \qquad \text{for all } v \in V .$$

3.3 - Let us choose a basis $m_0,\ldots,m_{\nu-1}$ of the K-linear subspace of K^{μ} defined by the equations

$$\sum_{i=0}^{\mu-1} a_{ij}x_i = 0,\ j = 0,\ldots,\mu-\nu-1 .$$

Let us fix Z by requiring that its columns take the values $m_0,\ldots,m_{\nu-1}$ respectively, at ξ .
Because of (3.2.1), we have $i_v(Y)(\xi) = (i_v(Z)(\xi))\cdot B_v$, for some suitable column B_v with ν entries in K_v , depending on $v \in V$. Because $i_v(Y)$ and $i_v(Z)$ are both solutions of Λ , hence uniquely determined by their "initial" value at ξ , we have

$$i_v(Y) = i_v(Z)B_v \text{ , identically.}$$

This implies that

$$i_v(r) = i_v(P.Z)B_v ,$$

from which we deduce

(3.3.1) $\mathrm{ord}_{\xi}\, i_v(r) \geqq \alpha$, for every $v \in V$.

3.4 - Let us consider, for $v \in V$, the v-adic analytic function $i_v(r)\, x^{-\beta}\left(1-\frac{x}{i_v(\xi)}\right)^{-\alpha}$ in $D_v(0,R_v(Y))$. It takes the value $i_v(\eta)$ at 0. Let $\varepsilon > 0$ be such that $|\xi|_v < r_v := \mathrm{Min}(1,R_v(Y))e^{-\varepsilon/|V|}$.
The maximum modulus principle in the disk $D(0,r_v)$ gives

$$\text{(3.4.1)}\quad \log|\eta|_v \leqq \mathop{\mathrm{Max}}_{i,n} \log^+|p_{i,n}|_v + \alpha \log|\xi|_v/\mathrm{Min}(1,R_v(Y))$$

$$+ \mathop{\mathrm{Max}}_{|x|_v \leqq r_v} \log\|Y(x)\|_v + \alpha \log|2|_v^{-1} + \beta \log^+ 1/R_v(Y)$$

$$+ \beta\, O(1)\varepsilon + o(\alpha) .$$

3.5 - We next sum our inequalities (3.4.1) over all $v \in V$ and (3.1.1) over all $v \notin V$; since $\eta \neq 0$, the "product formula" $\sum_v \log|\eta|_v$ holds and yields, after dividing by α :

$$\sum_{v\in V}\log|\xi|_v + (1+\deg P)\frac{h(P)}{\alpha} + \beta/\alpha\ \sigma(Y) + \left(\sum_{v\in V}\log^+1/R_v(Y)\right)(1+\beta/\alpha) + \varepsilon\ 0(1) + o_\varepsilon(1) \geq -\log 2$$

(where the last quantity may depend on ε). We now use 2.3., we use the previous lemma, and let $\alpha \to \infty$, then $\varepsilon \to 0$; we then get:

$$\sum_{v\in V}\log|\xi|_v + \left\{\frac{\nu}{\mu}(1+\tau)+\tau(2s-1)\right\}h(\xi)$$
$$+ \left\{1+s\tau+\frac{\nu}{\mu}(1+\tau)\right\}\log 2 + 2\tau\sum_{S\,\mathrm{in}\,\Lambda} h(\zeta) + \tau\ \sigma(\Lambda)$$
$$+ \frac{\nu(1+\tau)}{\tau}\sigma(Y) + \left(1+\frac{\nu(1+\tau)}{\tau}\right)\sum_{v\in V}\log^+1/R_v(Y) \geqq 0 .$$

Hence we have proven the following

PROPOSITION. Under the assumptions 2.1 and 3.2, we have the following constraint on ξ :

$$\sum_{v\in V}\log|\xi|_v+\left\{\frac{\nu}{\mu}(1+\tau) + \tau(2s-1)\right\}h(\xi) \geqq -c_4(\Lambda,\tau,Y) ,$$

where $c_4(\Lambda,\tau,Y) = (s\tau+\frac{\nu}{\mu}(1+\tau)+1)\log 2 + \tau(\sigma(\Lambda) + 2\sum_{S\,\mathrm{in}\,\Lambda} h(\zeta))$
$+ \frac{\nu(1+\tau)}{\tau}\sigma(Y) + (1+\frac{\nu(1+\tau)}{\tau})\rho(Y)$

(and the last two terms may be replaced by $(1+\frac{\nu(1+\tau)}{\tau})\sigma(Y)$ if $(h_{v,n}(Y))_n$ has a limit for every $v \in V$).

REMARK 1. $\rho(Y) < \infty$ follows from $\sigma(\Lambda) < \infty$ by the results of chapter IV and V (more precisely, we may express explicitely these two invariants in terms of $\sigma(Y),\mu,\Gamma)$. In practice however, it is more convenient to compute $\rho(Y)$ and $\rho(\Lambda)$; we have also given expressions for $\rho(Y)$ and $\sigma(Y)$ in terms of $\rho(\Lambda)$, see chapter V.

REMARK 2. The proposition is non-empty only if τ is such that $\frac{\nu}{\mu}(1+\tau) + \tau(2s-1) < 1$, i.e. $\tau < \frac{\mu-\nu}{(2s-1)\mu+\nu}$.

3.6 - APPLICATION: HILBERT's IRREDUCIBILITY THEOREM (after P. Bundschuh, V.G. Sprindzuk and P. Debes [20 1/2]. Let $p\in K[x,y]$, and assume that $p(0,y)$ has a simple root in K . Hence there exists an algebraic function $y_1 \in K[[x]]$ such that $p(x,y_1) \equiv 0$; moreover $Y = {}^t(y_0=1, y_1,\ldots,y_{\mu-1}=y_1^{\mu-1})$ has linearly independent entries over $K(x)$ if $\mu = \deg_y p$, and is a solution of a

G-operator Λ (the fact that $\sigma(Y) < \infty$ comes either from Eisenstein's theorem about algebraic functions, or proposition I 4.2). The reducibility of $p(\xi,y) \in K[y]$ for some $\xi \in K^*$ would yield dependence relations among the $y_i(\xi)$. Replacing ξ by ξ^m for m large, this contradicts under suitable assumptions the inequality displayed in the last proposition. We can get this way some effective versions of Hilbert's irreducibility theorem; a typical example is

COROLLARY (Sprindzuk [57]). Assume that the quantities $|\xi|_v$, where v runs over the set of places such that $|\xi|_v < 1$, constitute a non-empty family of multiplicatively independent real numbers. Then for any sufficiently large m (effective), $p(\xi^m,y)$ is irreducible in $\mathbf{K}[y]$.

§ 4. LOCAL DEPENDENCE

4.1 - In addition to the assumptions 2.1, we suppose that the entries of Y are homogeneously algebraically independent over $K(x)$, (hence $\mu \geq 2$). We then may replace Y by $Y^{\odot n}$ and Λ by $\Lambda^{\odot n}$ for any $n > 0$. The new invariants may be controlled as follows:

$$\mu_n = \binom{\mu+n-1}{\mu-1},$$

$$\sigma(Y^{\odot n}) \leq \sigma(Y)(1 + \log n),$$

$$\sigma(\Lambda^{\odot n}) \leq \sigma(\Lambda)(1 + \log n),$$

the global radii, and $\operatorname{Sin}\Lambda$, remain invariant.

Instead of 3.2, we suppose that there is some non-trivial homogeneous relation of degree δ among the $i_v(y_i(\xi))$'s, with coefficients in $i_v(K)$ for some place v such that $|\xi|_v \leq 2^{-d_v/d} \operatorname{Min}(1,R_v(Y))$. Multiplying this relation by monomials of degree $n-\delta\geq 0$ in the y_i's, we get $\binom{\mu+n-1-\delta}{\mu-1}$ linearly independent linear dependence relations among the $n_v(y_i(\xi))$'s. Hence we may take

$$\nu_n = \binom{\mu+n-1}{\mu-1} - \binom{\mu+n-1-\delta}{\mu-1} \leq \frac{\delta\sum_{i=0}^{\mu-2}\delta^i(n+\mu-1)^{\mu-i-2}\binom{\mu-1}{i}}{(\mu-1)!} \leq \frac{\delta(\delta+n+\mu-1)^{\mu-2}}{(\mu-2)!}$$

We choose $\tau = \dfrac{\delta}{n(2s-1)}$.

Then $\dfrac{\nu_n}{\mu_n}(1+\tau)+s\tau \leq \dfrac{\nu_n}{\mu_n}(1+\tau)+\tau(2s-1)$

$\leq (\mu-1)\frac{\delta}{n}\, 2^{\mu-1}$ if $n \geq \delta+\mu-1$,

and $\dfrac{\nu_n(1+\tau)}{\tau} + 1 \leq \dfrac{n^{\mu-1}}{(\mu-2)!}\, 2^{\mu-1}(2s-1)$. We assume that $h(\xi) \geq e$ and that

(4.1.1) $h(\xi)/\log h(\xi) \geq \mathrm{Max}(\sigma(Y)+\Sigma\, h(\zeta)+\log 2,\ \dfrac{(\delta+\mu)^{\mu}e^{\mu}}{\delta})$,

and we choose

(4.1.2) $n = \left[\left(\delta\, h(\xi)/\log h(\xi)\right)^{1/\mu}\right] + 1$.

Thus , if we make the mild hypothesis that s is the same as the one defined in VI 4,

$$(\mu-1)\frac{\delta}{n}\, 2^{\mu-1}\left\{h(\xi)+\log 2 + \frac{\sigma(\Lambda)(\log n+1)+2\ \Sigma h(\zeta)}{(2s-1)2^{\mu-1}(\mu-1)}\right\}$$

$$\leq (\mu-1)\frac{\delta}{n}\, 2^{\mu}h(\xi)$$

$$\leq 2^{\mu}(\mu-1)\delta^{1-1/\mu}h(\xi)^{1-1/\mu}(\log h(\xi))^{1/\mu}$$

(for the first inequality, we used the formula given in theorem VI 4: $\sigma(\Lambda) \leq 5\mu^2 s\ \sigma(Y)$ and half of (4.1.1)). Similarly, using again (4.1.1), we find

$$\frac{n^{\mu-1}}{(\mu-2)!}\, 2^{\mu-1}(2s-1)(\sigma(Y)(\log n+1) + \log^{+}1/R_v(Y))$$

$$\leq \frac{2^{2\mu+1}}{\mu!}(\mu-1)s\ \delta^{1-1/\mu}h(\xi)^{1-1/\mu}(\log h(\xi))^{1/\mu}\sigma(Y)$$

(here we also use the fact that $\log^{+}1/R_v(Y) \leq \sigma(Y)$). Putting everything together, we find that

<u>if</u> $h(\xi) \geq \mathrm{Max}(\sigma(Y)+\Sigma h(\zeta)+\log 2,\ (\delta+\mu)^{\mu}e^{\mu}/\delta)\ \mathrm{Max}(1,\log h(\xi))$

<u>and if</u>

$$|\xi|_v \leq \exp\left\{-2^{2(\mu+1)}\sigma(Y)\,\mathrm{Max}\left(\frac{s}{(\mu-1)!}\ \delta^{1-1/\mu}h(\xi)^{1-1/\mu}(\log h(\xi))^{1/\mu}, 1\right)\right\} ,$$

<u>then there cannot be any non-trivial homogeneous relation of degree</u> δ <u>among the elements</u> $y_0(\xi),\ldots,y_{\mu-1}(\xi)$ <u>of</u> K_v .

At this point, it is important to remind the normalization of $|\ |_v$. Note also that the case of inhomogeneous relations can be derived by replacing Y by $Y^{+} = \begin{pmatrix} Y \\ 1 \end{pmatrix}$ and Γ by $\Gamma^{+} = \begin{pmatrix} \Gamma & 0 \\ 0 & 0 \end{pmatrix}$.

4.2 - All of this generalizes without difficulty if we only assume that at least one of the y_i/y_0 is transcendental (instead of assuming that the transcendence degree of $K(x)((y_i/y_0)_i)$ is $\mu-1$). However it would be difficult to present an explicit statement as above.

The proof follows closely 4.1, so that we give only a brief sketch (see [7] 12 for more details).

Since the entries of $Y^{\otimes n}$ are no longer assumed linearly independent, one has to replace $\Lambda^{\otimes n}$ by a "sub-operator" $\Lambda^{(n)}$ defined in the following way: let M_n be the vector space over $K(x)$ generated by all monomials $y_{i_1},\dots,y_{i_n}$, $0\leq i_1\leq\dots\leq i_n\leq\mu-1$, and let $\mu_{(n)} := \dim_{K(x)} M_n$; let us fix $\mu_{(n)}$ monomials $y_{i_1}\dots y_{i_n}$ linearly independent over $K(x)$ and let $Y^{(n)}$ be the corresponding vector. Then there exists a unique matrix $\Gamma^{(n)}$ with entries in $K(x)$ such that $\Lambda^{(n)}Y^{(n)} := d/dx\, Y^{(n)}-\Gamma^{(n)}Y^{(n)}=0$.

From lemmata 2 in IV 3, IV 4, it follows that

$$\rho(\Lambda^{(n)}) \leq \rho(\Lambda^{\otimes n}) \leq \rho(\Lambda) ,$$

$$\sigma(\Lambda^{(n)}) \leq \sigma(\Lambda^{\otimes n}) \leq \sigma(\Lambda)(1+\log n)$$

$$(\leq 5\mu^2 s\, \sigma(Y)(1+\log n), \; s \text{ being defined}$$

according to VI 4), while we have

$$\mathrm{Sin}\Lambda^{(n)} \subseteq \mathrm{Sin}\Lambda^{\otimes n} (\subseteq \mathrm{Sin}\Lambda) ;$$

this follows from the following remark: if we write Y' for the complement of $Y^{(n)}$ in $Y^{\otimes n}$, so that $Y' = BY^{(n)}$ for a suitable $(\mu_n-\mu_{(n)}) \times \mu_{(n)}$ - matrix B with entries in $K(x)$, then any solution $X^{(n)}$ of $\Lambda^{(n)}$ lifts to a solution $\begin{pmatrix} X^{(n)} \\ BX^{(n)} \end{pmatrix}$ of $\Lambda^{\otimes n}$. At last, we have

$\mu_{(n)} = \dim H^0(\mathrm{Proj}\oplus M_m, \mathcal{O}(n)) = \chi(\mathrm{Proj}\oplus M_m, \mathcal{O}(n))$ for large n , in particular $\mu_{(n)} \sim \deg(\mathrm{Proj}\oplus M_m)\, \frac{n^{\mu'}}{\mu'!}$ as $n \to \infty$ where μ' stands for $\dim \mathrm{Proj}\oplus M_m = \mathrm{tr.deg}\, K(x)((y_i/y_0)_i)$.

Now we are faced to the problem of counting how many independent linear relations hold among the $i_v(y_{i_1}\dots y_{i_n})(\xi)$'s ; let us denote this number by $\mu_{(n)} - \nu_{(n)}$. Hence $\nu_{(n)} = \dim_K H^0(\mathrm{Proj}\oplus M_{m,\xi}, \mathcal{O}(n))$, where $M_{m,\xi}$ denotes the K-vector space generated by the monomials $i_v(y_{i_1}\cdots y_{i_m})(\xi)$. Following Bombieri, we shall say that there is a <u>strongly non-trivial homogeneous relation of degree</u> δ <u>among the</u> $y_i(\xi)$'s, if there is a hypersurface Σ_δ of degree δ in $\mathbb{P}_K^{\mu-1}$ such that

$$\mathrm{Proj}\oplus M_{m,\xi} \subseteq (\underset{x\to\xi}{\mathrm{Specializ.}}\ \mathrm{Proj}\oplus M_m) \cap \Sigma_\delta =: X_\xi ,$$

and such that this intersection X_ξ has dimension $<\mu'$ (hence $= \mu'-1$). The point is that a strongly non-trivial relation <u>decreases</u> the transcendence degree. Then

$$\begin{aligned}\nu_{(n)} &\leqq \dim_K H^0(X_\xi,\mathcal{O}(n)) = \chi(X_\xi,\mathcal{O}(n)) \quad \text{for } n >> 0 ,\\ &\sim \deg X_\xi \frac{n^{\dim X_\xi}}{(\dim X_\xi)!}\\ &\leqq \delta \deg(\mathrm{Proj}\oplus M_m)\cdot\frac{n^{\mu'-1}}{(\mu'-1)!}\end{aligned}$$

by Bezout's theorem (see [34] I 7.7). (This is a simplification of Bombieri's argument [7] 12). Using the semi-continuity of coherent cohomology applied to Chow variety which parametrizes the X_ξ's, one can show that these estimates are uniform in ξ.

4.3 - THEOREM. <u>Let</u> $y_0,\dots,y_{\mu-1}$ <u>be</u> μ <u>G-functions. Then there exist two constants</u> c_5 <u>and</u> c_6 <u>depending only on</u> $y_0,\dots,y_{\mu-1}$ <u>with the following property: let</u> δ <u>be a positive integer,</u> v <u>a place of</u> K, <u>and let</u> μ' <u>denote the homogeneous transcendence degree of the field generated by the</u> y_i's <u>and their derivatives over</u> $K(x)$; <u>if</u> $h(\xi) \geqq c_5\,\delta^{\mu'}(\log\delta+1)$ <u>and if</u>

$$|\xi|_v \leqq \exp(-c_6\,\delta^{1-1/\mu'+1}h(\xi)^{1-1/\mu'+1}\log h(\xi)^{1/\mu'+1}) ,$$

<u>then there cannot be any strongly non-trivial homogeneous relation of degree</u> δ <u>among the elements</u> $y_i(\xi)$ <u>of</u> K_v.

Proof: note first that if $\mu' = 0$, there cannot be any strongly non-trivial relation at all, hence we may assume that $\mu' \geqq 1$. If Λ_i is a matrix-operator associated with the scalar operator which kills y_i, we may construct $\Lambda = \oplus\Lambda_i$ and $Y = (y_0,y_0',y_0'',\dots,y_{\mu-1}^{(\mathrm{ord}(\Lambda_{\mu-1})-1)})$, so that $\Lambda Y=0$ and $\sigma(Y) < \infty$. Moreover μ' corresponds to the homogeneous transcendence degree of $K(x)(Y)$ over $K(x)$. Now we construct the differential operators $\Lambda^{(n)}$, which are uniquely determined by the choice of $\mu_{(n)}$ independent monomials of degree n in the y_i's. We assume that c_5 is large enough to ensure that

$$h(\xi) \geq c_5 \Rightarrow \xi \notin \mathrm{Sin}\Lambda \supseteq \mathrm{Sin}\Lambda^{(n)} .$$

Next we use our previous upper (resp. lower) bound for $\nu_{(n)}$ (resp. $\mu_{(n)}$). Finally, the rest of the argument goes as in the special (but explicit) case 4.1 but choosing

$$n \sim (\delta\ h(\xi)/\log h(\xi))^{1/\mu'+1} .$$

□

REMARK 1. In [7] 12, a similar result is stated, but with $h(\xi)^{1-\frac{1}{2\mu'}}$ instead of $h(\xi)^{1-1/\mu'+1}$. Moreover one imposes there a priori, that Λ is a G-operator, and that $\rho(Y) < \infty$; at last, the proof given loc.cit. uses Dwork-Robba's theorem, and a refined version of Siegel's lemma (we did not so).

REMARK 2. It turns out that $\mathrm{Proj}\oplus M_m$ and its specialization at ξ (at least for ξ ordinary) are torsors (= principal homogeneous space) under suitable differential Galois groups, see chapter IX. Hence if "the" differential Galois group is (geometrically) connected, these varieties are geometrically irreducible, and every non-trivial relation is automatically strongly non-trivial.

§ 5. GLOBAL DEPENDENCE; HASSE'S PRINCIPLE

5.1 - Let p be a homogeneous polynomial in μ variables, with coefficients in K , and let $Y = {}^t(y_0,\dots,y_{\mu-1}) \in K[[x]]^\mu$. We say that a relation $p(y_0(\xi),\dots,y_{\mu-1}(\xi)) = 0$ <u>holds</u> v-<u>adically</u> if $i_v(p)(i_v(y_0)(i_v(\xi)),\dots,i_v(y_{\mu-1})(i_v(\xi))) = 0$ in K_v ; such a relation is said <u>non-trivial</u> if it does not come by specialization at ξ from a homogeneous relation of the same degree (with coefficients in $K[x]$) among the series y_i . According to § 4, we call it <u>strongly non-trivial</u> if it does not occur as a factor of the specialization at ξ of a homogeneous irreducible relation among the series y_i (possibly of higher degree). Following Bombieri, we say that a relation $p(y_0(\xi),\dots,y_{\mu-1}(\xi)) = 0$ <u>is a global relation if it holds</u> v-<u>adically for every place</u> $v \in K$ <u>for which</u> $|\xi|_v < \mathrm{Min}(1,R_v(Y))$. In other words, a global relation is a relation which is true v-adically whenever it is defined. Let us remark that for a global relation, the fact that it is non-trivial (resp. non-trivial) at one place v where it is defined, is independent of this place v. In his paper of 1981 [7], E. Bombieri remarked for the first time that the existence of non-trivial global relations implies a finiteness property. In fact, the following result may be considered as a Hasse principle for values of G-functions.

5.2. THEOREM. Assume that $Y \in \bar{\mathbb{Q}}[[x]]$ satisfies the differential system $d/dx\, Y = \Gamma Y$ for some $\Gamma \in M_\mu(\bar{\mathbb{Q}}(x))$, and that $\sigma(Y) < \infty$. Let $Ш_\delta(Y)$ (resp. $Ш'_\delta(Y)$) denote the set of ordinary points or apparent singularities $\xi \in \bar{\mathbb{Q}}^x$ where there is some non-trivial (resp. strongly non-trivial) global (homogeneous) relation of degree δ. Then $h(Ш_\delta(Y)) \leq c_7(Y)\cdot\delta^{3(\mu-1)}(\log\delta+1)$

and $$h(Ш'_\delta(Y)) \leq c_8(Y)\cdot\delta^{\mu}(\log\delta+1) .$$
In particular, any subset of $Ш_\delta(Y)$, with bounded degree over $\mathbb{Q}$, is finite.

Proof: let V be the set of places $v \in \Sigma(K)$ such that

$$|\xi|_v < \mathrm{Min}(1,R_v(Y)) \quad \text{if} \quad v \in \Sigma_0$$
$$|2\xi|_v < \mathrm{Min}(1,R_v(Y)) \quad \text{if} \quad v \in \Sigma_\infty .$$

We have $$\sum_{v\in V}\log|\xi|_v = -\sum_{v\in V}\log^+|1/\xi|_v = -h(\xi)+\sum_{v\notin V}\log^+|1/\xi|$$
$$\leq -h(\xi) + \rho(Y) + \log 2 .$$

Looking at theorem 3.5, we see that any choice of τ such that $\tau < \frac{\mu_{(\delta)}-1}{2s\mu_{(\delta)}-1}$ will lead to an upper bound for $h(\xi)$, after replacing Λ by $\Lambda^{(\delta)}$). Since $\mu_{(\delta)} = O(\delta^{\mu-1})$, the required estimate for $h(Ш_\delta)$ will follow after choosing $\tau = \frac{1}{2(\mu+2s)}$ for instance.

In order to estimate $h(Ш'_\delta)$, we use $\Lambda^{(n)}$ instead of $\Lambda^{(\delta)}$, for appropriate large n (as in the previous theorem, whose proof carries over in the present case without an essential change.) The finiteness assertion follows from the bound for the height and Northcott's theorem, see e.g. [45]. □

5.3 REMARK. However, relations between special values of G-functions need not be "global", even in the particular case of rational values of algebraic functions at rational points. In order to understand

better this phenomenon, the reader is invited to carry out the case of the Pythagorean function $\sqrt{1+x^2}$. Of course, by multiplying all the "local" relations, obtained for each place v such that our algebraic functions converge at the given point ξ (there are only a finite number of such v), we obtain this way a global relation; but in most cases this relation comes from a functional (= "trivial") relation by specializing.
In this respect, the case of transcendental G-functions $y_0,\dots,y_{\mu-1}$ is easier to handle, because a product of strongly non-trivial relations remains strongly non-trivial; thus in practice, it suffices to construct, for each place v such that $y_0(\xi),\dots,y_{\mu-1}(\xi)$ has a sense v-adically, a strongly non-trivial relation with algebraic coefficients which links these v-adic values $y_0(\xi),\dots,y_{\mu-1}(\xi)$.

§ 6. APPLICATION TO DIOPHANTINE EQUATIONS

6.1 We now apply the "Hasse principle" which has just been proved, to the arithmetics of superelliptic curves, and obtain that certain rational or algebraic points (with some conditions about denominators) are in finite number. Moreover we shall display quite explicit and fairly "small" bounds. We also take this opportunity to refine our prop. 3.5 in the case of algebraic functions, and show how to compute the invariants in this concrete situation.

6.2 Two diophantine results

Let us consider an irreducible superelliptic curve C over $\mathbb{Z}$ defined by the equation

$$(6.2.1) \quad y^m = q(z) ,$$

where q denotes a monic polynomial with coefficients in $\mathbb{Z}$, of degree say n.
We assume moreover that

(6.2.2) m and n have a prime common factor $l \geq 3$, or else are divisible by $l = 4$. Let us denote by $H(q)$ the maximum among $\pm$ the coefficients of q, and let us set $H(\xi) = \exp h(\xi)$ (if $\xi = a/b \in \mathbb{Q}$ is an irreducible ratio, this is simply $\mathrm{Max}(|a|,|b|)$).

THEOREM 1. *There are only finitely many rational points* (y,z)

on C such that no prime $\equiv 1 \bmod. l$ divides the denominator of z. For such a point, one has the bound

$$H(z) < (2^{8n} l^3 H(q)^2)^{3n+2} .$$

THEOREM 2. There are only finitely many totally real points in $C(\bar{\mathbb{Q}})$ with bounded denominator and degree.

6.3 Proof. 1) Let us consider the column Y consisting in the two G-functions $y_0 = 1$ and $y_1 = (x^n q(1/x))^{2/l}$. Hence $Y(0) = \begin{pmatrix} 1 \\ 1 \end{pmatrix}$, and Y satisfies the differential system of order $\mu = 2$:

$$\Lambda Y := d/dx\, Y - \begin{pmatrix} 0 & 0 \\ 0 & \frac{2x^{-n}}{l\, q(1/x)} \end{pmatrix} Y = 0$$

One has $\operatorname{Sin} \Lambda = \{$reciprocals of the non-zero roots of $q\}$, and $|\operatorname{Sin} \Lambda| = s \leq n$.

Now, let us compute the invariants of Y. Since the components of Y are globally bounded (Eisenstein, or direct checking on the expansion below), one has $\sigma(Y) \leq \rho(Y) = \rho(y_1)$.

Let us write $q = x^{\alpha} \prod_{\zeta} (x - 1/\zeta)$, so that $y_1 = \prod (1 - x/\zeta)^{2/l}$. By looking at the expansion of $(1 - \frac{x}{\zeta})^{2/l} = \sum_{j \geq 0} (-2/l)_j (\zeta^{-j}/j!) x^j$, one finds (I app.):

$$\rho(y_1) \leq \sum_{p \mid l/2} (v_p(l/2) + \frac{1}{p-1}) \log p + h(\operatorname{Sin} \Lambda)$$

$$\leq 3/2 \log l/2\, v_2(l) + (v_2(l/2)+1) \log 2 + h(\operatorname{Sin} \Lambda)$$

$$\leq 3/2 \log l + h(\operatorname{Sin} \Lambda) .$$

But $h(\operatorname{Sin} \Lambda) \leq \sum_{\zeta \text{ s.t. } q(1/\zeta)=0} h(1/\zeta)$ (since $h(\zeta) = h(1/\zeta)$)

$= \log \prod M(1/\zeta)^{1/\deg \zeta}$ (with the help of the Mahler measure M, see I 1.1)

$= \log \prod \operatorname{Max}(1, |1/\zeta|)$ since q has rational coefficients,

$\leq n \log 2 + \log H(q)$ by a well-known inequality.

Therefore

(6.3.1) $\rho(Y) \leqq n \log 2 + 3/2 \log l + \log H(q)$.

Similarly the size of the matrix Z involved in the proof of theorem 3.5 can be estimated, because

$$Z = \begin{pmatrix} 1 & 1 \\ \prod(1-\frac{x-\xi}{\zeta-\xi})^{2/l} & 0 \end{pmatrix},$$

and therefore

$$\sigma(Z) \leqq \rho(Z) \leqq \sum_{\zeta\in \mathrm{Sin}\Lambda} h(\frac{1}{\zeta-\xi}) + 3/2 \log l$$

(6.3.2) $\begin{cases} \leqq 2n \log 2 + 3/2 \log l + \log H(q) + s\, h(\xi) \\ \text{for an ordinary point } \xi \in \bar{\mathbb{Q}} . \end{cases}$

2) Let us now write down the relations which occur in case there exists a point $P = (y,z)$ as in the theorem. Let us set $\xi = 1/z$.
Then any singularity of Λ gives rise to a point P on $C(\bar{\mathbb{Q}})$, but they are in finite number and certainly satisfy the required inequality for the height.
Hence we assume from now onwards that $\xi \notin \mathrm{Sin}\,\Lambda$. Let $K = \mathbb{Q}(y,z)$ denote the field of definition of P . We shall show that the following relation between the values of y_0, y_1 at $\xi = 1/z$:

(6.3.3) $(y^2 z^{-2n/l})y_0(\xi) - y_1(\xi)$,

is a global (resp. "almost" global) non-trivial relation if $K = \mathbb{Q}$ and $|\xi|_p \geqq 1$ for every prime $p \equiv 1(l)$ (resp. if K is totally real of bounded degree and $h_f(1/\xi)$ is bounded), which corresponds to the hypothesis of theorem 1 (resp. theorem 2). Let us notice that (6.3.3) is non-trivial because y_0 and y_1 are linearly independent over $\bar{\mathbb{Q}}(x)$; indeed $l > 2$, and C is assumed to be irreducible, so that q is neither a lth nor a $(l/2)^{th}$ power.
On the other hand, by putting $y_2 = (x^n q(1/x))^{1/l} = \sqrt{y_1}$, (6.3.3) can be written in the following form:

(6.3.4) $\prod_{\varepsilon=\pm 1} (y_2(\xi) - \varepsilon y \xi^{n/l}) = 0$

(recall that $l|n$).

But the following identity of formal power series:

$$y_2^{\,l} = x^n q(1/x) ,$$

implies that for any place v such that

$$|\xi|_v < \mathrm{Min}(1,R_v(Y)) ,$$

one has $y_2^{\,l}(\xi) = y^l \xi^n$ v-adically, i.e.

$$\prod_{\varepsilon^l=1}(y_2(\xi) - \varepsilon y \xi^{n/l}) = 0 .$$

Thereby relation (6.3.4) holds true whenever for such a place v,

(6.3.5) the only l^{th} roots of unity in K_v are ± 1.

The point is simply that $y_2(\xi) \in K_v$.

In the case of theorem 1, every place v of $\mathbb{Q}$ such that $|\xi|_v < 1$ satisfies $v = \infty$ or $v = p \not\equiv 1(l)$, hence (6.3.5) is fulfilled. In the case of theorem 2, every infinite place v of K satisfies (6.3.5), because $K_v = \mathbb{R}$. Since moreover $h_f(1/\xi)$ is bounded, a slight modification of theorem 5.2 shows that $h(\xi)$ itself is bounded; all the same in the case of theorem 1. And because $[K:\mathbb{Q}]$ is bounded too, there is only a finite number of points P in both cases.

3) Under the hypothesis of theorem 1, we shall push further the computations, using (6.3.2) and avoiding the general proposition IV 5.4. Let V denote the set of all places v of $\mathbb{Q}$ such that $\mathrm{Min}(1,R_v(Y)) > |\xi|_v$ (resp. $|2\xi|_v$ if $v = \infty$).

In terms of the approximating forms, the argument of § 3 revisited gives rise to the inequality

$$\sum_{v\in V}\log|\xi|_v + \underline{\lim}_{\alpha\to\infty}(1+\deg P_{(\alpha)})\frac{h(P_{(\alpha)})}{\alpha} + \nu(1+\tfrac{1}{\tau})\overline{\lim_{j\to\infty}}\sum_{v\notin V} h_{v,j}(Y)$$

$$+\Big(\sum_{v\in V}\log^+ 1/R_v(Y)\Big)(1+\nu(1+1/\tau)) + \log 2 \geq 0 .$$

On the other hand, we have

$$\overline{\lim_{\alpha\to 0}}(1+\deg P_{(\alpha)})\frac{h(P_{(\alpha)})}{\alpha} \leq \frac{\nu}{\mu}(1+\tau)(h(\xi)+\log 2) + \tau\sigma(Z)$$

(see lemma 2.2).

In our present case, $\mu = 2$, $\nu = 1$, and Y is globally bounded. Replacing $\rho(Y)$ and $\sigma(Z)$ by the bounds (6.3.1) and (6.3.2) respectively, we find

$$\sum_{v\in V}\log|\xi|_v+\frac{1+\tau}{2}(h(\xi)+\log 2)+\log 2+\tau(2n\log 2+3/2\log l+s\,h(\xi)+\log H(q))$$

$$+(2+\frac{1}{\tau})(n\log 2+3/2\log l+\log H(q))\geq 0\ ,$$

hence

$$\frac{1-\tau(1+2s)}{2}h(\xi)\leq\frac{\log 2}{2}\cdot(7+4n+\tau(1+4n)+\frac{2n}{\tau})+\log(l^{3/2}H(q))\cdot(2+\tau+\frac{1}{\tau})\ .$$

We next choose $\tau=\frac{1}{4s+2}$. A straightforward computation then gives the required inequality

$$h(\xi)<(4s+6)(4n\log 2+3/2\log l+\log H(q))\ ,$$

and we conclude by noting that $h(\xi)=\log H(z)$. It is also true that $H(z)<10^{10n^2}H(q)^{8n}$. □

REMARKS. 1) This bound should be compared with Baker's upper bound for integral solutions of the "general" superelliptic equation: $y^m=q(z)$, $m\geq 3$, $q\in\mathbb{Z}[z]$ with at least two simple roots; then for integral (y,z) ,

$$\mathrm{Max}(|y|,|z|)<\exp\exp[(5m)^{10}n^{10n^3}H(q)^{n^2}]$$

cf. A. Baker, "bounds for the solutions of the hyperelliptic equation", Proc. Cambridge Phil. Soc. 65(1969) 439-444.

2) In his paper "On Weil's "théorème de décomposition"" Am.J.Math. 105 n°2 (1983) 295-308, E. Bombieri has pointed out that the algebraic function case of a variant of theorem 3.5 could be obtained by a geometric method, using Weil functions and the quadraticity of the Néron-Tate height.

EXERCISES. 1) Prove a statement analogous to theorem 5.2 for globally bounded series $y_i\in K[[x]]$ without assuming that the y_i's satisfy any differential equation, but with a bound for the height of $\xi\in\underline{|||}_\delta$ of the form $c(Y,q)$, which may depend on the height of the relation q too. (Hint: use the fact that ξ corresponds to a pole of the v-adic series $i_v(q(y_0,\ldots,y_{\mu-1}))^{-1})$.

2) From 5.2 deduce the full "theorem E" in the introduction.

3) (Siegel's method). Hypotheses are as in 2.1 and 3.2. Let N be a sufficiently large integer and let $0<\tau<1$. Construct

first μ polynomials $p_0,\dots,p_{\mu-1} \in K[x]$ such that

$$\operatorname{Max} \deg p_j \leqq N$$

$$\operatorname{ord}_0\Big(\sum_{l=0}^{\mu-1} p_l y_l\Big) \geqq (\mu-\tau)N$$

$$h(\text{coef. } p_0,\dots,p_{\mu-1}) \leqq \frac{(\mu-\tau)^2}{2\tau}\,\sigma(Y)N + o(N) \ .$$

Recall from VI 1 (3) the sequences of rational functions

$$_{nj}R = \sum_{h=0}^{n} \sum_{l=0}^{\mu-1} \frac{1}{h!}\Big(\frac{d^h}{dx^h} p_l\Big)\,{}_{lj}\Gamma_{n-h} \ .$$

Show that $\det({}_{nj}R)_{j,n=0,\dots,\mu-1} \neq 0$, using V 1 (5), the fuchsianity of Λ and III, appendix, or else a variant of V 2 (Shidlovski).

Let $\sum_{i=0}^{\mu-1} a_{ij} y_i(\xi) = 0,\ j = 0,\dots,\mu-\nu-1$ denote relations which hold v-adically for v belonging to a subset V of the set of all places such that $|\xi|_v < \operatorname{Min}(1,R_v(Y))$. Construct a non-zero determinant

$$\Delta = \begin{vmatrix} a_{ij} \\ {}_{n_i,j}R(\xi) \end{vmatrix} \neq 0 \ , \text{ with } \ n_i \leqq \tau N + \text{cst.} \ ;$$

for $v \in V$, one has $\Delta y_j(\xi) = \sum_{i=0}^{\nu} (\text{cofactor } {}_{\mu-\nu+i,j}\Delta) \times \frac{1}{n_i!} \frac{d^{n_i}}{dx^{n_i}}\Big(\sum_{l=0}^{\mu-1} p_l y_l\Big)$. Use this equality to bound $\sum_{v\in V} \log|\Delta|_v$ from above. Using the product formula for Δ, and letting $N \to \infty$, deduce a statement analogous to theorem 3.5 (by controling $h({}_{n_i,j}R(\xi))$ in terms of $\sigma(\Lambda)$, $h(P)$, $h(\xi)$), see [7] 11.

4) Show that if we consider only integral points in theorem 1, the hypothesis $l \geqq 3$ can be lowered: $l \geqq 2$ is enough; but "q monic" is essential (think at the Pell-Fermat equation).

APPENDIX: ON RUNGE'S METHOD

Runge's argument, which is exactly one-century old and perhaps a little forgotten, was the first general method for solving diophantine equations in two variables. We shall describe it in the special case treated in § 6 in order to emphasize the similarity between this method and the G-function method used in this chapter;

E. Bombieri, loc.cit., already pointed out that Runge's theorem [52] could be easily obtained from G-function theory.
On the other hand, non-ramification hypotheses of the type (6.2.2) are typical features from both methods; the next case not covered by either of them is the Thue equation.
Let us consider a solution $(y,z) \in \mathbb{Z}^2$ of the irreducible equation $y^m = q(z)$, such that m and $n = \deg q$ have a common factor $l \geq 2$ (not necessarily prime).
Let us consider the algebraic function

$$g(x) = q(1/x)^{1/l} = x^{-n/l} + \dots + g_{n/l} + g_{n/l+1}x + \dots ,$$

and let us choose l polynomials $p_0,\dots,p_{l-1} \in \mathbb{Z}[x]$ of degree $\leq n/l$ and not all zero, such that

$\mathrm{ord}_0\left(\sum_0^{l-1} p_j(1/x)g^j(1/x)\right) > 0$. This is possible, because this linear system has more unknowns $(l(n/l+1))$ than equations $(n+1)$. For a sufficiently large solution z, if any, and for the value $\xi = 1/z$, $\sum_0^{l-1} p_j(1/\xi)g^j(1/\xi) = \sum_0^{l-1} p_j(z)y^{mj/l} \in \mathbb{Z}$, but $\sum_0^{l-1} p_j(1/\xi)g^j(1/\xi) = h\xi + \text{h.o.t.} \ll 1$, hence vanishes. Now y is a common root of the two polynomials $\sum_0^{l-1} p_j(z)y^{mj/l}$ and $y^m - q(z)$, and we find that z lies among the finitely many roots of the resultant polynomial.
Clearly it is possible to give effective bounds for (y,z) this way, and the method also extends to the case of rational points with bounded denominator.

Chapter VIII A Criterium of Rationality

§ 1. STATEMENT OF THE RESULTS

1.1. In this logically independent chapter, we present various generalizations of the Borel-Dwork criterion; all of them are derived from a single "main criterion" whose proof relies on the diophantine method already used in the last chapter, namely Gel'fond's method at several places.

Let us first recall the statement of the classical BOREL-DWORK CRITERION
Let $y \in K[[x]]$ (K = a number field). Then y is a rational function if and only if it is globally bounded and $\prod_{v \in \Sigma} M_v(y) > 1$.

Here $M_v(y)$ stands for the v-adic radius of meromorphy of y , that is to say, the greatest element $r \in [0,\infty]$ such that $i_v(y) := \Sigma\, i_v(y_n)x^n \in K_v[[x]]$ extends to a meromorphic function on the disk $D(0,r)$ of $\mathbb{C}_v$ (this definition extends obviously to series in several variables, replacing disks by polydisks). A particular case of this criterion states that any globally bounded series $y \in K[[x]]$ which satisfies $\prod_v R_v(y) > 1$ is a polynomial.

Our first generalization deals with the problem of weakening the assumption of global boundedness (which however cannot be omitted, e.g. think at $y = \Sigma \frac{x^n}{n!}$ with $M_\infty(y) = \infty$). We prove:

THEOREM. Let $y \in K[[x_1,\ldots,x_\nu]]$. Then y is a rational function iff $\sigma(y) < \infty$ and $\prod_v M_v(y) > 1$.

1.2. Can one give a criterion of algebraicity along similar lines? The answer turns out to be yes: it suffices to replace the meromorphy condition on y by the assumption that both y and x are uniformized (meromorphically).

Let $\| \ \|_v$ denote the "sup norm" on either $\mathbb{C}_v^\nu$ or $M_\nu(\mathbb{C}_v)$.

THEOREM. Let $y \in K[[x_1,\ldots,x_\nu]]$. Assume that for each place v , y , resp. $x_1,\ldots,x_\nu$ can be expressed as meromorphic functions $y_v(z_1,\ldots,z_\nu)$, resp. $x_{1,v}(z_1,\ldots,z_\nu),\ldots,$ on the polydisk $\|\underline{z}\|_v < M_v$ of $\mathbb{C}_v$, with the following "normalizations":

i) $\underline{x}(\underline{0}) = \underline{0}$,

ii) the Jacobian matrix $\left(\frac{D\underline{x}_v}{D\underline{z}}\right)_{|\underline{0}}$ is invertible,

and iii) $\left\| \left(\frac{D\underline{x}_v}{D\underline{z}}\right)^{-1}_{|\underline{0}} \right\|_v \leq 1$.

If y is globally bounded and $\prod_v M_v > \nu^2$, then y is algebraic over $K(\underline{x})$.

REMARKS. 1) The normalization iii) rules out dilatations in $\underline{z}$ which could trivially enlarge the M_v's .

2) The assumption $\sigma(y) < \infty$ would not suffice here, as the following example shows: $y = \sum_{n\geq 0} \frac{x^n}{n} = z$ at $v = \infty$, $x = 1-e^{-z}$, $M_\infty = \infty$.

But one could replace the global boundedness by the (a priori) weaker assumption $\tilde{\sigma}(y) := \operatorname{Sup}_n \sigma(y,y^2,\ldots,y^n) < \infty$, if one replaces at the same time $\prod M_v > \nu^2$ by $\prod M_v > \nu^2 \cdot e^{\tilde{\sigma}}$.
As a corollary of the theorem refined in this way, we have:

COROLLARY. Let $y \in K[[\underline{x}]]$ with $\tilde{\sigma}(y) < \infty$. Choose an embedding $K \subset \mathbb{C}$, and assume that y and the x_i's are uniformized by meromorphic functions of $\underline{z}$ on $\mathbb{C}^\nu$, such that the Jacobian determinant $\left|\frac{D\underline{x}}{D\underline{z}}\right|$ does not vanish identically. Then y is algebraic over $K(\underline{x})$.

□

REMARK. For sufficient conditions which imply $\tilde{\sigma}(y) < \infty$, see ex.3 below.

REMARK-CONSTRUCTION . In fact the uniformization (sufficient) condition of theorem 2 is also necessary up to adding some new variables, at least if one starts with $\nu = 1$.

Indeed, let $y \in K\{x\}$ be an algebraic function (in one variable). Dividing by a monomial, we may assume that $y(0) = 1$. Let us consider the complete smooth curve, say C, over K associated with the function field $K(x,y)$, and let D denote the divisor of poles of the differential form $\omega_1 = ydx$ on C.
Let $\omega_1, \dots, \omega_\nu$ be a basis of $\Omega^1_C(-D)$, and let us write J_D for the generalized Jacobian of C associated with the "module" D. For every complex embedding ι_v of K, we get a sequence of meromorphic maps:

$$\begin{array}{ccccccc} \mathbb{C}^\nu \simeq (\Omega^1_C(-D))^* & \longrightarrow & (\Omega^1_C(-D))^* \big/ H_1(C\setminus \mathrm{Supp}\, D) & \xrightarrow{\sim} & J_D & \longrightarrow & C^{(\nu)} \\ \Psi & & & & & & \Psi \\ \left(z_i = \sum_j \int_0^{x_j} \omega_i\right) & & \longmapsto & & & & \sigma_j(x_i) \end{array}$$

where $C^{(\nu)}$ denote the ν-fold symmetric product of C with itself, (ν = genus $C - 1 + \mathrm{Max}(1, \deg D)$). Thus $y = \frac{\partial z_1}{\partial \sigma_1}\Big|_{\sigma_2 = \dots = \sigma_\nu = 0}$ can be expressed in terms of the (meromorphic) jacobian matrix $\frac{D\underline{x}}{D\underline{z}}$, and y and $\sigma_j(x_i)$ are simultaneously uniformized by meromorphic functions on $\mathbb{C}^\nu$, (of exponential order of growth ≤ 2 after Serre [54] 2/3) In other words, one can choose $M_v = \infty$ for any $v \in \Sigma_\infty(K)$ in the theorem.

1.3. Now remember the classical

POLYA CRITERION. Let $y \in K[[x]]$. Then y is rational iff y is globally bounded and $\frac{dy}{dx}$ is rational.

As a direct consequence of the above corollary and construction (applied to $\frac{dy}{dx}$, we obtain the following variant:

COROLLARY (of the corollary). Let $y \in K[[x]]$. Then y is algebraic iff y is globally bounded and $\frac{dy}{dx}$ is algebraic.[1]

□

(The fact that an algebraic function is globally bounded is Eisenstein's theorem, see I § 4.2.).

(1) This statement was suggested to me by J.P. Bézivin.

1.4. If one is willing to weaken the global boundness assumption in theorem 1.2., one has to put some extra growth condition upon the meromorphic uniformization.
We shall say that a quotient f/g of two entire functions on $\mathbb{C}^{\nu}$ has (exponential) order of growth $\leq \gamma$ if $\max_{\|\underline{z}\| \leq \kappa} (|f(\underline{z})|, |g(\underline{z})|) \leq$ $\leq \exp A\kappa^{\gamma}$ for some constant A, and any sufficiently large κ (here the sup norm $\| \ \|$ is relative to the Euclidean metric on $\mathbb{C}$).

THEOREM (Chudnovsky-Chudnovsky). Let $Y = (y_0,\ldots,y_{\mu-1}) \in$ $\in K[[x_1,\ldots,x_\nu]]^{\mu}$. Assume that for each infinite place v, there is a simultaneous uniformization on $\mathbb{C}_v = \mathbb{C}$ as in theorem 1.2 (with $M_v = \infty$), except that condition iii) is replaced by:

iv) the $x_{i,v}$ have order of growth $\leq \gamma < \infty$.

If moreover $\overline{\lim}_n \frac{\sigma(Y^{\leq \Theta n})}{\log n} < \mu/_{\gamma\nu}$, then $y_0,\ldots,y_{\mu-1}$ are algebraically dependent over $K(\underline{x})$ (here $Y^{\leq \Theta n}$ denotes the set of degree $\leq n$ in the y_i's).

EXAMPLE. Recall our notation $L_1 = -\log(1-x)$. Let us prove that $\overline{\lim}_{n\to\infty} \sigma(L_1^n)/_{\log n} = 1$, as was announced in I 4.3. Indeed, by lemma 2g) of chapter I 1.4, one has $\overline{\lim}_{n\to\infty} \sigma(L_1^n)/_{\log n} \leq \sigma(L_1) = 1$. On the other hand $\sigma(L_1^n) = \sigma(L_1^n,\ldots,\frac{(n-k)!}{n!}(-(1-x)d/dx)^k L_1^n,\ldots,1)$

$$= \sigma(1,L_1,\ldots,L_1^n) .$$

One may uniformize x and L_1 simultaneously by putting $x = e^{-z}$, which has order of growth 1, $L_1 = z$. Since L_1 is transcendental, corollary 4 applies and gives the reversed inequality $\overline{\lim}_{n\to\infty} \sigma(L_1^n)/_{\log n} = \overline{\lim}_{n\to\infty} \sigma((1,L_1)^{\Theta n})/_{\log n} \geq 1$.

□

REMARKS. 1) Actually it is required in [19] that the uniformizing functions y_v's also have order of growth $\leq \gamma$, but this turns out to be unnecessary.

2) If one combines this theorem with the construction 1.2, one may modify corollary 1.3 (variant of the Polya criterion) in the following fashion: replace "globally bounded" by $\overline{\lim_n} \frac{\sigma(1,y,y^2,\ldots,y^n)}{\log n} < 1/2\nu$. Examples to be given below show that this is best possible in general.

1.5 Topological interlude. The following examples, arising from the theory of elliptic genera (see [47]), are intended to showing that the estimate in theorem 1.4. is rather sharp and sensitive. Recall that a power series $q(z) = 1+a_2z^2 + a_4z^4 + \ldots \in \mathbb{C}[[z^2]]$ determines a homomorphism (genus) $\varphi : \Omega \to \mathbb{C}$, where Ω is the Thom cobordism ring of oriented differentiable manifolds: writing formally the total Pontrjagin class of a 4k-dimensional manifold X as $(1+z_1^2) \ldots (1+z_{2k}^2)$, then $\varphi(X) = q(z_1) \ldots q(z_{2k})$. Each such genus has a logarithm:

$$y = \int_0^x \sum_{n\geq 0} \varphi(\mathbb{C}\,\mathbb{P}^{2n})\, t^{2n} dt \in x\mathbb{C}[[x^2]] .$$

It is related to the characteristic series q in the following fashion: taking $y = z$ as "uniformizing parameter", then $q(z) = y/x$. A genus φ is called elliptic if

$$y = \int_0^x (1-2\delta t^2+\varepsilon t^4)^{-1/2} dt \quad \text{for some } \delta, \varepsilon \in \mathbb{C} .$$

We note that if $\delta,\varepsilon \in \bar{\mathbb{Q}}$, y is a G-function; it satisfies the differential equation of order two:

$$(*) \quad (1-2\delta x^2 + \varepsilon x^4) \frac{d^2y}{dx^2} + 2x(\varepsilon x^2 - \delta) \frac{dy}{dx} = 0$$

In fact, if $\varepsilon \neq 0$, there is the expansion

$y = \sum_{n\geq 0} P_n(\delta/\sqrt{\varepsilon})\, \varepsilon^{n/2} \frac{x^{2n+1}}{2n+1}$, where the P_n's denote the Legendre polynomials. For instance, if $\delta = \varepsilon = 1$, one has

$y = \int_0^x \frac{dt}{1-t^2} = \sum_{n\geq 0} \frac{x^{2n+1}}{2n+1}$, and $q(z) = z/\tanh z$ is the characteristic series of the signature (L-genus). And if $\delta = -1/8$, $\varepsilon = 0$, one has $y = \int_0^x \frac{dt}{\sqrt{1+t^2/4}} = \sum_{n\geq 0} (-1)^n \binom{2n}{n} 2^{-4n} \frac{x^{2n+1}}{2n+1}$, and $q(z) = z/2\,\mathrm{sh}\, z/2$.

is the characteristic series of the $\hat{A}$-genus. In both cases, $x(z) = z/q(z)$ is an entire function of order of growth $\gamma = 1$. It is easily seen on the expansions that at least the prime-to-2 part of the size is ≤ 1; it follows that $\overline{\lim_n}\, \sigma((1,y)^{\odot n})/_{\log n} \leq 1$ in both cases. As Argth and Argsh are transcendental, we see that the estimate in theorem 1.4 is best possible in both cases. Now, let us assume that $\varepsilon(\delta^2-\varepsilon) \neq 0$. Then y is an elliptic logarithm in the usual sense, and $x(z)$ is a Jacobi "sn" function, which is meromorphic of order of growth $\gamma = 2$. We next assume that $\delta,\varepsilon \in \bar{\mathbb{Q}}$ and that the elliptic curve $E_{\delta,\varepsilon} : x'^2 = \varepsilon x^4 - 2\delta x^2 + 1$ has complex multiplication. Then from the fact that $E_{\delta,\varepsilon}$ mod. v is supersingular for a set of places v of density 1/2, it can be deduced that $\overline{\lim_n}\, \sigma((1,y)^{\odot n})/_{\log n} \leq 1/2$, because supersingular places v induce vanishing v-curvatures for the differential equation (*) (see ex. 4 for more details). As an elliptic logarithm is transcendental, we see once again that the estimate in theorem 1.4 is best possible in this case.

At last, it is perhaps worth pointing out that this connexion with algebraic topology is no artifice, but the integrality and congruence properties of the coefficients of the elliptic logarithm lies contrariwise at the root of theory of elliptic cohomology, already in the proof of its existence, see [47].

1.6. All the above results are subsumed in the following

MAIN CRITERION. Let $Y = (y_0,\ldots,y_{\mu-1}) \in K[[x_1,\ldots,x_\nu]]^\mu$, let $\tau > 0$, and let $V \subset \Sigma(k)$ be some set of places of K. Assume that for each $v \in V$ the y_i's and x_j's are simultaneously uniformized by (v-adic) meromorphic functions on a polydisk of $\mathbb{C}_v^\nu$ of radius $> \kappa_v$, with the "normalizations" i), ii), iii) (of theorem 1.2). Write $x_{j,v} = \frac{f_{j,v}}{g_v}$ as a quotient of analytic functions with $g_v(\underline{0}) = 1$, and set

$$\chi_v = \log \sup_{|\underline{z}|_v=\kappa_v} (|f_{j,v}|,|g_v|) .$$

If the following inequality holds

(*) $\sigma_{\text{not } V}(Y) + \tau\,\sigma(Y) + 2\log\nu < \sum_{v\in V}\left[\log\kappa_v - \left(\frac{1}{\mu}\left(1+\frac{1}{\tau}\right)\right)^{1/\nu}\chi_v\right]$,

then the y_i's are linearly dependent over $K(\underline{x})$. Moreover if $\underline{x}_v = \underline{z}$ or if $V \subset \Sigma_f(K)$, the term $2\log\nu$ may be omitted.

(Here $\sigma_{\text{not } V}$ is defined like a size but the summation runs only over those v not in V; hence, if $\Sigma\setminus V$ is finite,
$\sigma_{\text{not } V}(Y) = \sum_{v \text{ not in } V} \log^+(1/R_v(Y))$.

REMARK. Inequality (*) can be satisfied only for $\tau > \frac{1}{\mu-1}$, when $\underline{x}_v = \underline{z}$.

COROLLARY. Let $y \in K[[x]]$. Then y is rational iff $e^{12\sigma(y)} < \prod_{v\in V} M_v(y)$ for some (resp. any) subset V of $\Sigma(K)$.

Indeed, take $y_0=1$, $y_1=y$, $\nu=1$, $\mu=2$, $x=z$, $\chi_v = \log\kappa_v$, $\kappa_v \longrightarrow M_v^-$ in the previous criterion, Then (*) is satisfied for $\tau = 2$ ($\tau = \sqrt{2}+1$ gives a slight improvement : 11,66 instead of 12).

□

In particular if $\sigma(y) = 0$ and if sime $M_v(y) > 1$, then y is rational with poles in the set of roots of unity (since for a rational function, $\sigma(y) = h(\text{pol } y)$, see I 4.1). Can one replace the assumption $M_v(y) > 1$ by "y is a G-function" in this statement?

1.7. Deduction of theorems 1.1, 1.2, 1.4.

1.7.1. We shall prove theorem 1.1 in two steps: at first we show that y is algebraic. We set $Y = (1,y,\dots,y^{n-1})$, so that $\sigma(Y) \leq (1+\log(n-1))\sigma(y)$. Let us set $\mu = n$ and $\tau = \frac{1}{\sqrt{n}}$, $\underline{x}_v = \underline{z}$.
For n large enough, inequality (*) is then satisfied with $V = \Sigma(K)$, if the κ_v are chosen closed enough to the M_v's. Hence y is algebraic.
Multiplying y by a polynomial, we may assume that $R_v = M_v$ (elimination of the apparent singularities of the ∇-module associated with y). Moreover $R_v=1$ for almost all v, because y is a diagonal of a rational function. Thus it is enough to show that any series $y \in K[[\underline{x}]]$ satisfying

a) $R_v(y) = 1$ for almost all v

b) $\prod R_v(y) > 1$

is a rational function (in fact a polynomial). Given an integer $N > 0$ and a multiindex $\underline{m}$ with $\mathrm{Max}\ m_i < N$, and given $\zeta \in K^\times$, we form the series $\psi_{\underline{m}}\ y = \sum_{\underline{n}} \zeta^{|\underline{n}|} y_{N\underline{n}+\underline{m}} \underline{x}^{\underline{n}}$. The v-adic radius of this series is given by $R_v(\psi_{\underline{m}}\ y) = (R_v(y))^N / |\zeta|_v$; therefore, by the product formula, $\psi_{\underline{m}}\ y$ still satisfies a) and b).

Let S denote the set of places either infinite or such that $R_v(y) \neq 1$. Applying the pigeon-hole principle to the lattice given by the S-unit theorem and the $N'+1$ points $(R_v(y)^n)_{v\in S}$, $n = 0,\dots,N'$, we find N, $0 < N \leqslant N'$, and some S-unit ζ such that $e^{-c_v/N'} \leq \dfrac{R_v(y)^N}{|\zeta|_v} \leq e^{c_v/N'}$ for any $v \in S$ except one, and for constants c_v independent of N'. By choosing N' sufficiently large, we may assume that

$$\log \prod_{v\in S} \mathrm{Max}(1, R_v(\psi_{\underline{m}}y)^{-1}) < \frac{1-(3/4)^{1/\nu}}{2} \log \prod_{v\in S} R_v(\psi_{\underline{m}}y) .$$

Since $\sigma(\psi_{\underline{m}}y) \leq \rho(\psi_{\underline{m}}y) = \log \prod_{v\in S} \mathrm{Max}(1, R_v(\psi_{\underline{m}}y))$, inequality (*) (without the term $2 \log \nu$) in the main criterion is then satisfied for $y_0=1$, $y_1 = \psi_{\underline{m}}y$, $\mu = \tau = 2$ and $\underline{x} = \underline{z}$: hence the $\psi_{\underline{m}}y$'s are rational. Since $y(\zeta\underline{x}) = \sum_{\mathrm{Max}\ m_i<N} \psi_{\underline{m}}(y)\underline{x}^{\underline{m}}$, we find that $y(\zeta\underline{x})$, whence y itself, is rational. □

1.7.2. Proof of theorem 1.2.: since $\rho(y) < \infty$ and $\prod M_v > \nu^2$, one may choose $\varepsilon > 0$, a finite set V of places of K, and real numbers $0 < \kappa_v < M_v$ (for each $v \in V$) subject to the inequalities

$$\begin{cases} \sum_V \log \kappa_v - 2 \log \nu > \varepsilon , \\ \sum_{v\notin V} \log^+ 1/R_v(y) < \varepsilon/3 . \end{cases}$$

Let us choose $Y = (1, y, \dots, y^{\mu-1})$ in the main criterion, and $\tau < \varepsilon/_{3\sigma(y)}$. Because $\tilde{\sigma}(y) < \infty$, $\sigma(Y)$ and the left-hand side of (*) are bounded independently of μ. The same thing holds for χ_v. Let

μ be so large that $\sum_V \left(\frac{1+\tau}{\mu\tau}\right)^{1/\nu} \chi_v < \varepsilon/3$. Inequality (*) is then satisfied because $\sigma_{\text{not } V}(Y) \leq \rho_{\text{not } V}(y) := \sum_{v \notin V} \log^+ 1/R_v(y)$ for globally bounded y . Hence y is algebraic.

□

1.7.3. The proof of theorem 1.4 goes as follows: apply the main criterion for V = place associated with $K \hookrightarrow \mathbb{C}$, $Y = Y^{\odot \leq n}$, $\mu_n = \binom{\mu+n}{n} \sim n^\mu$ instead of μ . By assumption $\chi_v \leq \frac{d_v}{d} \kappa_v^\nu + \text{cst}$ for any $v \in \Sigma_\infty(K)$. Let us take $\kappa_v = n^{\mu/\gamma\nu}$.
By assumption, there exists $\varepsilon > 0$ such that $\sigma(Y^{\odot \leq n}) < (1-\varepsilon)(\mu/\gamma\nu)\log$ for every large enough n . For sufficiently large n (resp. small τ inequality (*) is thus fulfilled, whence again the conclusion that the y_i's are algebraically dependent.

□

§ 2. APPROXIMATING FORMS

Assume that $Y \in K[[\underline{x}]]^\mu$ satisfies $\sigma(Y) < \infty$. Recall that if $y_i = \sum_{\underline{n}} y_{i,\underline{n}} \underline{x}^{\underline{n}}$, then

$$\sigma(Y) := \overline{\lim_{m\to\infty}} \frac{1}{m} \sum_v \max_{\substack{|\underline{n}|\leq m \\ i=0,\dots,\mu-1}} \log^+ |y_{i,\underline{n}}|_v .$$

LEMMA. Let α be a positive integer, and let $\tau \in \mathbb{R}^+$. Then there exists $P = (p_0,\dots,p_{\mu-1}) \in \mathcal{O}_K[\underline{x}]^\mu$, with the following properties:

i) $N := \deg P \leq \left(\frac{1}{\mu}\left(1+\frac{1}{\tau}\right)\right)^{1/\nu} \alpha + o(\alpha)$, $P \neq 0$

ii) $h(\text{coef } P) \leq \tau\, \alpha\, \sigma(Y) + o(\alpha)$

iii) $\text{ord}_0 P.Y \geq \alpha$.

Proof: we write $p_i = \sum_{|\underline{n}|\leq N} p_{i,\underline{n}} \underline{x}^{\underline{n}}$. Then condition iii) is the following linear system of $\binom{\alpha+\nu-1}{\nu}$ equations in the $\mu\binom{N+\nu}{\nu}$ unknown quantities $p_{i,\underline{n}}$:

$$\sum_{i=0}^{\mu-1} \sum_{\substack{\underline{n}+\underline{m}=\underline{l} \\ |\underline{n}|\leq N}} p_{i,\underline{n}}\, y_{i,\underline{m}} = 0 \quad \text{for } |\underline{l}| < \alpha .$$

Since $\binom{N+\nu}{\nu} \sim N^{\nu}/_{\nu!}$ and $\binom{\alpha+\nu-1}{\nu} \sim \alpha^{\nu}/_{\nu!}$, this system has a non-zero solution if $N \sim \left(\frac{1}{\mu}\left(\left(1+\frac{1}{\tau}\right)\right)\right)^{1/\nu} \alpha$ for any $\tau > 0$. Furthermore, Siegel's lemma gives such a solution in $\mathcal{O}_K^{\mu\binom{N+\nu}{\nu}}$, with

$$h\left((p_{i,\underline{n}})_{i,\underline{n}}\right) \leq \frac{\binom{\alpha+\nu-1}{\nu}}{\mu\binom{N+\nu}{\nu}-\binom{\alpha+\nu-1}{\nu}} \left[h(y_{i,\underline{m}})_{\substack{i=0,\dots,\mu-1 \\ |\underline{m}| < \alpha}} + \log \mu\binom{N+\nu}{\nu} + \text{const.}\right]$$

$$\leq \tau\,\alpha\,\sigma(Y) + \circ(\alpha) \ .$$

□

In the sequel, we denote by r the series $P.Y = \sum p_i y_i$. In case $r \neq 0$, we denote by β the order of r at 0. Let us consider some coefficient of r of order β :

$$(2.1) \qquad \eta := \frac{1}{\underline{n}!} \left.\frac{\partial^{\underline{n}} r}{\partial \underline{x}^{\underline{n}}}\right|_{\underline{x}=\underline{0}} \qquad \text{with } |\underline{n}| = \beta \text{, such that } \eta \neq 0 \ .$$

We have written $\underline{n}!$ for $\prod_{i=1}^{\nu} n_i!$, if $\underline{n} = (n_1,\dots,n_\nu)$. We have $\eta \in K^{\times}$. One finds easily the following bound for the non-V part of the height of η :

$$(2.2) \qquad \sum_{v \notin V} \log^{+}|\eta|_v \leq h_{\text{not } V}(\text{coef } P) + \beta\sigma_{\text{not } V}(Y) + o(\beta)$$

§ 3. A METHOD OF GEL'FOND

3.1 Let us write $y_{i,v} = h_{i,v}/e_v$ as a quotient of two analytic functions, and let us put, in accordance with the hypotheses of the main criterion and the previous lemma,

$$\Psi_v(\underline{z}) = g^N_v(\underline{z}) \cdot \sum_{i=0}^{\mu-1} p_i(\underline{f}_v(\underline{z})/g_v(\underline{z})) \cdot h_{i,v}(\underline{z})$$

so that $\psi_v(\underline{z})$ is holomorphic in a neighbourhood of the polydisk $\|\ \underline{z}\ \|_v \leq \kappa_v$, and $\psi_v(\underline{z}) = g_v^N(\underline{z}) \cdot e_v(\underline{z}) \cdot r(\underline{f}_v(\underline{z})/g_v(\underline{z}))$ whenever $\|\ \underline{f}(\underline{z})/g(\underline{z})\ \|_v < R_v(Y)$

Moreover $|g_v(0)| \geq 1$, and we can also assume that $|e_v(\underline{0})| \geq 1$. The subscript v is present to recall that there exists such a holomorphic function ψ_v at every place of V .

3.2 Using the estimate for $f_{j,v}, g_v$ on $\| \underline{z} \|_v = \kappa_v$, we obtain

$$(3.2.1)\ \log|\psi_v(\underline{z})|_v \leq \frac{d_v}{d}\log \mu\binom{N+\nu}{\nu} + N\chi_v + \log^+ \operatorname*{Max}_{i=0,\dots,\mu-1} |h_{i,v}(\underline{z}), e_v(\underline{z})$$

$$+ h_v(\text{coef } P) .$$

We now apply Cauchy's integral formula in the polydisk $\| \underline{z} \|_v \leq \kappa_v$ to the holomorphic function ψ_v when $v \in V_\infty$:

$$\frac{1}{\underline{m}!} \frac{\partial^{\underline{m}}}{\partial \underline{x}^{\underline{m}}} \psi_v \Big|_{\underline{z}=\underline{0}} = \frac{1}{(2i\pi)^\nu} \oint_{\|\underline{z}\|_v=\kappa_v} \frac{\psi_v(\underline{t})\, dt_1 \dots dt_\nu}{\underline{t}^{\underline{m}}} .$$

Therefore:

$$(3.2.2)\quad \log\left|\frac{1}{\underline{m}!} \frac{\partial^{\underline{m}}}{\partial \underline{x}^{\underline{m}}} \psi_v \Big|_{\underline{z}=\underline{0}}\right|_v \leq \operatorname*{Max}_{\|\underline{z}\|_v=\kappa_v} \log|\psi_v(\underline{z})|_v - |\underline{m}| \log \kappa_v .$$

If $v \in V \cap \Sigma_f$, the same inequality holds (generalization in several variables of IV (1.5.1).)

For $|\underline{m}| = \beta$, we find:

$$(3.2.3)\ \sum_{v \in V} \log\left|\frac{1}{\underline{m}!} \frac{\partial^{\underline{m}}}{\partial \underline{x}^{\underline{m}}} \psi_v \Big|_{\underline{z}=\underline{0}}\right|_v \leq N\sum_V \chi_v + h_v(\text{coef } P) - \beta\sum_V \log \kappa_v + o(\beta)$$

$$\leq h_V(\text{coef } P) + \beta\sum_V \left(\chi_\nu\left(\frac{1+\tau}{\mu\tau}\right)^{1/\nu} - \log \kappa_v\right) + o(\beta$$

by using the lemma(part i)).

(3.3) We can now express η (or rather $i_v(\eta)$) in terms of the quantities $\frac{1}{\underline{m}!} \frac{\partial^{\underline{m}}}{\partial \underline{x}^{\underline{m}}} \psi_v \Big|_{\underline{z}=\underline{0}}$ for $|\underline{m}| = \beta$, using the fact that

$\frac{\partial^{\underline{m}}}{\partial \underline{x}^{\underline{m}}} \psi_v \Big|_{\underline{z}=\underline{0}}$ as well as $\frac{\partial^{\underline{m}} r}{\partial \underline{x}^{\underline{m}}}$ vanish for $|\underline{m}| = \beta$: because

$\psi_v(\underline{z})/g_v^N e_v$ "is" $r(\underline{x})$, we find

$$(3.3.1)\quad \underline{\eta}!\ i_v(\eta) = i_v\left(\frac{\partial}{\partial x_{i_1}} \cdots \frac{\partial}{\partial x_{i_\beta}}\Big|_{\underline{x}=\underline{0}}\right)$$

$$= (g_v^{-N} e_v^{-1})(\underline{0}) \sum_{j_m=1}^{\nu} \left(\prod_{m=1}^{\beta} \left(\frac{D\underline{x}}{D\underline{z}}\Big|_{\underline{0}}\right)^{-1}_{i_m, j_m}\right) \frac{\partial}{\partial x_{j_1}} \cdots \frac{\partial}{\partial x_{j_m}} \psi_v \Big|_{\underline{z}=\underline{0}} .$$

By our normalization of g_v, e_v, $\frac{D\underline{x}}{D\underline{z}}$ at $\underline{0}$, the combination of last displayed formula and (3.2.3) gives:

$$(3.3.2)\qquad \sum_{v\in V} \log|\eta|_v \leq 2\beta \log \nu + \sum_{v\in V} \log \operatorname*{Max}_{|\underline{m}|=\beta} \left| \frac{1}{\underline{m}!} \frac{\partial^{\underline{m}}}{\partial \underline{x}^{\underline{m}}} \psi_v \Big|_{\underline{z}=\underline{0}} \right|_v$$

$$\leq h_V(\text{coef } P) + \beta\Big(\log \nu^2 + \sum_V \Big(\chi_v\Big(\frac{1+\tau}{\mu\tau}\Big)^{1/\nu} - \log \kappa_v\Big)\Big) + o(\beta$$

3.4 Using inequality (2.2), we can now conclude. Since $\eta \neq 0$, the "product formula" yields :

$$(3.4.1)\quad 0 = \sum_{v\notin V} \log|\eta|_v + \sum_{v\in V} \log|\eta|_v \leq h(\text{coef } P) + \beta\Big(\log \nu^2 + \sum_V \Big(\chi_v \Big(\frac{1+\tau}{\mu\tau}\Big)^{1/\nu} - \log \kappa_v$$

$$+ \sigma_{\text{not } V}(Y)\Big) + o(\beta)$$

while $h(\text{coef } P) \leq \tau\, \alpha\sigma(Y) + o(\alpha)$.

Note that if $\underline{z} = \underline{x}_v$ or $V \subset \Sigma_f$, the term $\log \nu^2$ arising in (3.3.2) can be omitted.

For α sufficiently large, these inequalities are in contradiction with the assumption (*) of the main criterion, since $\beta \geq \alpha$. Hence for some $\alpha \in \mathbb{N}$, the linear form in $y_0,\ldots,y_{\mu-1}$ with coefficients in $K(x)$ (not all 0) defined by r vanishes.

□

§ 4. APPLICATION: THE ISOGENY THEOREM, AFTER CHUDNOVSKY

The "isogeny theorem" states that if two elliptic curves defined over $\mathbb{Q}$ have the same number of points modulo "sufficiently many" primes, they are isogeneous over $\mathbb{Q}$. This is a consequence of Falting's solution of Tate's conjecture, completed by the work of Serre ["Quelques applications du théorème de Chebotarev", Publ. Math. IHES 54 (1981)] p. 196. Building upon their criterium of algebraicity D.V. and G.V. Chudnovsky have established a new and simple proof of a result of this type [19 1/2], namely:

THEOREM (Chudnovsky). Let E_1 and E_2 be two elliptic curves defined over $\mathbb{Q}$, and let $H(E_i)$ be the inverse of the area of a fundamental domain of the period lattice Γ_{E_i} of E_i with respect to a Néron differential form on a minimal model of E_i over $\mathbb{Z}$, for $i = 1,2$. If E_1 and E_2 have the same number of points modulo each prime $p \le c_\varepsilon(H(E_1)H(E_2))^{1+\varepsilon}$, then E_1 and E_2 are isogeneous over $\mathbb{Q}$.

Here ε denotes any positive number and C_ε is an effective constant depending only on ε.

Since a detailed exposition of this result already exists [46], we shall be very brief about the proof.

Assume first that E_1 and E_2 have the same number of points modulo every prime. It then follows from the theory of one-dimensional formal group that there is a commutative diagram

$$(**)\qquad \begin{array}{ccccc} & \overset{\exp E_1}{\nearrow} & \mathbb{C}/\Gamma_{E_1} & \xrightarrow{\ r_1\ } & \mathbb{C}\mathbb{P}^1 \\ \mathbb{C} & & & & \Big\uparrow y \in \mathbb{Z}[[x_1]] \\ & \underset{\exp E_2}{\searrow} & \mathbb{C}/\Gamma_{E_2} & \xrightarrow[\ r_2\ ]{} & \mathbb{C}\mathbb{P}^1 \end{array}$$

, where r_1, r_2 are non-constant rational functions on E_1, E_2. Note that $x_i = r_i \circ \exp_{E_i}(z)$, $i=1,2$, is meromorphic (or order of growth 2). Since obviously $\sup_n \sigma(1,y,\dots,y^n) < \infty$, it follows from theorem 1.2 or 1.4 that y is algebraic over $\mathbb{Q}(x_1)$, that is, there exists $q \in \mathbb{Z}[x_1,x_2]$ such that $q(r_1\circ\exp_{E_1}, r_2\circ\exp_{E_2}) = 0$. This implies that E_1 and E_2 are isogeneous.

In the general case, one obtains $y \in \mathbb{Q}[[x_1]]$ with good p-integrality properties only for $p \le C_\varepsilon(H(E_1)H(E_2))^{1+\varepsilon}$. The proof is carried over on building upon the following idea: in the course of proving the criterium of rationality, we use only the β first coefficients of y, for such a β that (3.4.1) contradict the growth assumption on the first coefficients of y; it remains to exhibit explicitly the $o(1)$'s which appear in (3.4.1) respectively; see [46] for a detailed account.

EXERCISES. 1) Show that $\sigma(x/L_1) = \infty$, as was announced in I 1.5. (Hint: show that $\sigma(1,x/L_1,\dots,(x/L_1)^n) \le \sigma(x/L_1)$ on using the successives derivatives of x/L_1).

For the following excercises, recall from IV ex 2,3,4, the formalism of v-curvatures of a scalar differential operator Λ in the Weyl algebra $A_1(\mathcal{O}_K)$.

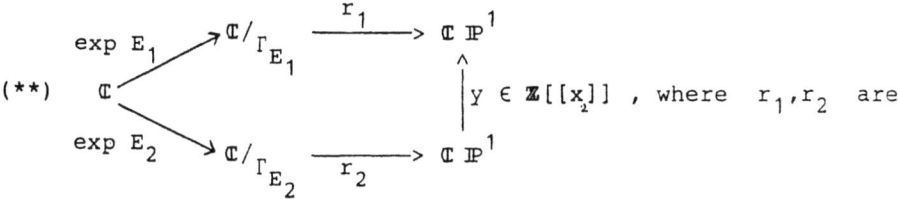

2) Let $\partial - G$ denote the matrix-valued differential operator of order one associated with Λ . Show that $\| G_{[n]} \|_v \geq |p|_v^{[n/p]}$ for all n , if the v-curvature ψ_v of Λ vanishes.

3) Deduce that the size of a solution $y \in K[[x]]$ of Λ is finite if $\psi_v = 0$ for almost every v , at least if 0 is an ordinary point. Show moreover that $\sigma(1,y,\ldots,y^n)$ is bounded independently from n in this case. (Do not use the Dwork-Robba theorem nor Frobenius structures ...). Extend this to quotients of solutions.

4) Using the Dwork-Robba theorem, prove the following generalization of Ex.3: let Λ be a G-operator of order μ in $A_1(\mathcal{O}_K)$ such that $\psi_v(\Lambda) = 0$ for a set of places v which has Dirichlet density δ , and such that 0 is ordinary; then for any solution of Λ in $K[[x]]$, $\overline{\lim\limits_{n\to\infty}}\, \sigma(1,y,\ldots,y^n)/_{\log n} \leq \mu(1-\delta)$. In particular, consider the G-operator (*) in § 1.5, associated to an elliptic curve with complex multiplication; show that its v-curvatures vanish for a set of places v of density $\delta = 1/2$ (hint: see IV ex.5, and [44] appendix 2).

5) (after Chudnovsky). Let M be a ∂-module of rank 1 over an affine curve C defined over a number field K ; denoting by x a local parameter on C, M is given by a differential equation $d/dx\, m = fm$ for some $f \in \mathcal{O}_C$. Prove the <u>Grothendieck conjecture</u> in this case, which asserts that if M has non-trivial "rational" solution modulo almost every finite place v of K , then M is solvable by means of algebraic functions. (Hint: first show that the hypothesis implies the vanishing of almost every v-curvatures of a differential equation $\Lambda \in A_1(\mathcal{O}_K)$ associated with the direct image of M on the x-line, on using IV ex.3.

Remark that a solution of M is given by $e^{\int f dx}$, and deduce that $y = \exp(\sum_{j=1}^{\nu} \int^{x_j} f dx) \in K[[x_1,\ldots,x_\nu]]$ satisfies

$\overline{\lim\limits_{n\to\infty}}\, \sigma(1,y,\ldots,y^n)/_{\log n} = 0$. Choose ν as in remark 1.2.2.)

6) Deduce from ex.5 that any rank-one module with connection arising from geometry becomes trivial over an étale finite covering of the base, at least if the base is a curve defined over $\bar{\mathbb{Q}}$; compare with II ex.1. (Hint: use the fact that connections arising from geometry are "globally nilpotent", see [39] or V, appendix).

Part Four

G-Functions in Arithmetic Algebraic Geometry

Chapter IX Towards Grothendieck's Conjecture on Periods of Algebraic Manifolds

Let $\bar{\mathbb{Q}}$ denote from now onwards the algebraic closure of $\mathbb{Q}$ in $\mathbb{C}$. The Grothendieck conjecture predicts that polynomial relations with coefficients in $\bar{\mathbb{Q}}$ among the periods of an (algebraic) projective manifold X defined over $\bar{\mathbb{Q}}$ is determined by the Hodge cycles on the powers of $X_{\bar{\mathbb{Q}}}$ (or by the algebraic cycles, in the strongest version). Building upon methods of chapter VII and of variation of Hodge structure, we give a partial answer to this conjecture in general setting (§ 5).

§ 1. PERIODS

1.1 Definition. Let k be an algebraically closed subfield of $\mathbb{C}$. We consider a proper smooth variety X over k. Then there exists a canonical isomorphism

$$(1.1.1) \quad P_X : H^{\cdot}_{DR}(X) \otimes_k \mathbb{C} \longrightarrow H^{\cdot}(X^{an}_{\mathbb{C}}, \mathbb{Q}) \otimes_{\mathbb{Q}} \mathbb{C}$$

which relates the finite-dimensional $\mathbb{N}$-graded k-vector space (or complex with vanishing differential) of algebraic De Rham cohomology $H^{\cdot}_{DR}(X) = \mathbb{H}^{\cdot}(\Omega^{\cdot}_X)$, and the $\mathbb{N}$-graded $\mathbb{Q}$-vector space of cohomology of the constant sheaf $\mathbb{Q}$ on the analytic manifold $X^{an}_{\mathbb{C}}$ associated with $X \times_k \mathbb{C}$. This isomorphism comes as follows. There exists a spectral sequence

(1.1.2) $E_1^{pq} = H^q(X, \Omega^p_X) \Rightarrow \mathbb{H}^{\cdot}(\Omega^{\cdot}_X)$, known as the Hodge-De Rham spectral sequence (actually it degenerates at E_1, see [24 1/2]

for a simple algebraic proof). After base change $k \hookrightarrow \mathbb{C}$, one knows by GAGA [54 1/3] that the cohomology of any of the coherent sheaves $\Omega^p_{X_\mathbb{C}}$, thereby the hypercohomology of $\Omega^\cdot_{X_\mathbb{C}}$, are the same in the algebraic sense (Zariski topology) or the transcendental sense (usual topology on $X^{an}_\mathbb{C}$), because X is proper and smooth. In particular, $\Omega^{\cdot an}_{X_\mathbb{C}}$ being a resolution of the constant sheaf $\mathbb{C}$ by Poincaré lemma, one has a composed isomorphism

$$\begin{aligned}\mathbb{H}^\cdot(\Omega^\cdot_X) \otimes_k \mathbb{C} &= \mathbb{H}^\cdot(\Omega^\cdot_{X_\mathbb{C}}) \quad \text{by base change,}\\ &= H^\cdot(X^{an}_\mathbb{C},\mathbb{C})\\ &= H^\cdot(X^{an}_\mathbb{C},\mathbb{Q}) \otimes_\mathbb{Q} \mathbb{C} ,\end{aligned}$$

which is P_X , see [32]. On choosing a k-basis of $H^\cdot_{DR}(X)$, and a $\mathbb{Q}$-basis of $H^\cdot(X^{an}_\mathbb{C},\mathbb{Q})$, we may express P_X by a matrix with complex entries (in general transcendental over k). We call such a matrix, after multiplying by $(2i\pi)^{-n}$ in degree n , a <u>period matrix</u> of X ; an entry of the period matrix in degree n (i.e. at the level of the $n^{\underline{th}}$ groups of cohomology) will be called a <u>n-period</u> of X . We shall denote by $k(P_X{}^n)$, or $k(P^n)$ when no ambiguity follows, the extension of k inside $\mathbb{C}$ generated by the n-periods of X ; this field does not depend on any choice of bases.

EXAMPLE 1. Assume that X is a curve or an Abelian variety. Then

$H^1_{DR}(X) = \dfrac{\text{differentials of the second kind}}{\text{exact differentials}}$, and $H^1(X^{an}_\mathbb{C},\mathbb{Q})$ is dual to the space of rational 1-cycles in the singular homology of $X^{an}_\mathbb{C}$; the isomorphism P_X is just given by integration on the cycles, and any 1-period has the form $\frac{1}{2i\pi}\int_\gamma \omega$.

By analogy with this particular case, we shall denote n-periods in general by the symbols $\frac{1}{(2i\pi)^n}\int_\gamma \omega$ for $\omega \in H^n_{DR}(X)$, $\gamma \in H_n(X^{an}_\mathbb{C},\mathbb{Q}) := \mathrm{Hom}(H^n(X^{an}_\mathbb{C},\mathbb{Q}),\mathbb{Q})$, because $X^{an}_\mathbb{C}$ is compact).

EXAMPLE 2: diagonals of rational functions.

Let us first note that the isomorphism (1.1.1) holds even if X is not proper. The problem is to show that the hypercohomologies of $\Omega^\cdot_{X_\mathbb{C}}$ in the algebraic sense and in the transcendental sense coincide. To this aim, one constructs a smooth compactification

of $X_{\mathbb{C}}$, say $j : X \hookrightarrow \bar{X}$, such that the boundary $Y = \bar{X} \setminus X$ is a divisor with normal crossings (Hironaka). It is known that the complexes $j_*\Omega^{\cdot}_X$ and $\Omega^{\cdot}_X\langle Y\rangle$ are then quasi-isomorphic. One conludes by using a variant of the spectral sequence (1.1.2) with logarithmic singularities along Y and invoking GAGA. Hence periods can be defined as before.
In particular, let $p,q \in k[x_1,\dots,x_\nu]$, and let X be the $(\nu-1)$-dimensional affine smooth k-variety $k[x_1,\dots,x_\nu,\frac{1}{q}]/(x_1,\dots,x_\nu-\xi)$ for some $\xi \in k$. Then the period of the class of $p/q \cdot \frac{dx_2 \dots dx_\nu}{x_2 \dots x_\nu} \in H^{\nu-1}_{DR}(X)$ along the "vanishing" cycle $\gamma : |x_2| = \dots = |x_\nu| = \varepsilon$ (small) in $H_{\nu-1}(X^{an}_{\mathbb{C}},\mathbb{Z})$ is

$$\frac{1}{(2i\pi)^{\nu-1}} \int_\gamma p/q \cdot \frac{dx_2 \dots dx_\nu}{x_2 \dots x_\nu} = \Delta_\nu(p/q)(\xi)$$

, the diagonal of p/q evaluated at ξ, when $|\xi|$ is sufficiently small.

REMARK. The normalization by the factor $(2i\pi)^{-n}$ in degree n is somewhat arbitrary, but at least compatible with Shimura's definition of periods of CM-type [G. Shimura, "On the derivatives of theta functions and modular forms" Duke Math. J. 44 (1977) 365-387], and fits with the previous example. Anyway, we shall show in § 2.1 that for projective X, $(2i\pi)^n \in k(P^n_X)$, so that the introduction of this normalizing factor is innocuous.

1.2 The relative case. Assume now that S is a smooth (connected) k-variety, and let $f : X \longrightarrow S$ be a proper and smooth morphism. The relative De Rham cohomology $H^{\cdot}_{DR}(X/S) = \mathbb{R}^{\cdot} f_* \Omega^{\cdot}_{X/S}$ is a locally free graded $\mathcal{O}_S$-module of finite rank; on the other hand, the sheaf $R^{\cdot} f^{an}_* \mathbb{Q}_{X^{an}_{\mathbb{C}}}$ on $S^{an}_{\mathbb{C}}$ is a local system. The canonical isomorphism P_X has the following relative avatar:

$$(1.2.1) \quad P_{X/S} : H^{\cdot}_{DR}(X/S) \otimes_{\mathcal{O}_S} \mathcal{O}_{S^{an}_{\mathbb{C}}} \xrightarrow{\sim} R^{\cdot} f^{an}_* \mathbb{Q}_{X^{an}_{\mathbb{C}}} \otimes_{\mathbb{Q}_{S^{an}_{\mathbb{C}}}} \mathcal{O}_{S^{an}_{\mathbb{C}}} .$$

On choosing a basis of sections of $H^{\cdot}_{DR}(X/S)$ over an affine open subset U of S, and a trivialization (i.e. a frame) of $R^{\cdot} f^{an}_* \mathbb{Q}_{X^{an}_{\mathbb{C}}}$ over an analytic open subset V of $U^{an}_{\mathbb{C}}$, $P_{X/S}$ is expressed by a matrix with coefficients in $\mathcal{O}_V$. As before, we define the <u>relative period matrix</u> by multiplying this matrix by $(2i\pi)^{-n}$ in degree n . Using analytic continuation, it is easy to check that the field $k(P^n_{X/S})$ generated over $k(S)$ by the <u>relative n-periods</u> (i.e. the coefficients of the relative period matrix in degree n) is independent (up to isomorphism) of the choice of bases and of U, V . Recall that $H^n_{DR}(X/S)$ is the $\mathcal{O}_S$-module underlying an object of $\underline{\mathrm{MIC}}_{X/S}$ (see chapter II), namely $R^n f^{DR}_*(\mathcal{O}_X, d)$; in other terms, $H^n_{DR}(X/S)$ is endowed with the Gauss-Manin connection, denoted by ∇ . It gives rise to an analytic connection on $H^n_{DR}(X/S) \otimes_{\mathcal{O}_S} \mathcal{O}_{S^{an}_{\mathbb{C}}} \simeq \mathbb{R}^n f^{an}_* \Omega^{\cdot}_{X^{an}_{\mathbb{C}}/S^{an}_{\mathbb{C}}}$ (GAGA) , still denoted by ∇ . Then the canonical morphism of sheaves on $S^{an}_{\mathbb{C}}$

$$(1.2.2) \quad R^n f^{an}_* \mathbb{C} \xrightarrow{\sim} \mathbb{R}^n f^{an}_* (\Omega^{\cdot}_{X^{an}_{\mathbb{C}}}) \longrightarrow \mathbb{R}^n f^{an}_* \Omega^{\cdot}_{X^{an}_{\mathbb{C}}/S^{an}_{\mathbb{C}}}$$

is an isomorphism between its source and the subsheaf $\left(H^n_{DR}(X/S) \otimes_{\mathcal{O}_S} \mathcal{O}_{S^{an}_{\mathbb{C}}}\right)^{\nabla}$ of germs of horizontal sections of the target,

[40] 4.1.2. ...

This shows that the relative n-period matrix $\left(\frac{1}{(2i\pi)^n} \int_{\gamma_j} \omega_i\right)_{i,j=1,\ldots,\mu}$

is a complete solution in $GL_\mu(k(P^n_{X/S})) \subset GL_\mu(\mathcal{O}_V)$ of the Picard-Fuchs differential system of order one associated with $R^n f^{DR}_*(\mathcal{O}_X, d) = (H^n_{DR}(X/S), \nabla)$, endowed with the local basis $\omega_1, \ldots, \omega_\mu$.

REMARK. The isomorphism $P_{X/S}$ keeps existing when f is no longer proper but admits a good compactification $X \overset{j}{\hookrightarrow} \bar{X}$ over S (see ch. II;

by Hironaka's resolution of singularities, this happens if one replaces S by an étale neighbourhood). In addition to the GAGA argument applied to the fibres of $\bar{X}/S$, one has to use the fact that the complexes of $\mathcal{O}_{\bar{X}}$ - modules $\Omega^{\cdot}_{\bar{X}/S}\langle \bar{X} \setminus X\rangle$ and $j_* \Omega^{\cdot}_{X/S}$ are quasi-isomorphic (Atiyah-Hodge).

(1.2.3) EXAMPLE. Let us consider the elliptic surface $X/S = \bar{\mathbb{Q}}\,\mathbb{P}^1 \setminus \{0,1,\infty\}$ over the punctured J-axis, whose generic fibre X_η is

$$Y^2Z = 4X^3 - \frac{27J}{J-1}XZ^2 - \frac{27J}{J-1}Z^3 .$$

The functional invariant is J. Note that the fibre at $J = \infty$ minus its singular (double) point $(-3/2,0,1)$ inherits a group structure $\sim \mathbb{G}_m$, and $\left.\frac{dX}{2Y}\right|_{J=\infty}$ is regular away from the double point.

We set $x = 1/J$, and we are going to elucidate the periods in a neighbourhood of $J = \infty$, according to [59] 111-123. The differential equation (in the variable x) which annihilates the periods of $\omega_1 = \frac{dX}{2Y}$ is

$$\frac{d^2y}{dx^2} + \frac{1}{x}\frac{dy}{dx} + \frac{\frac{31}{144}-\frac{1}{36}x}{x(x-1)^2}\,y = 0 .$$

It is then possible to find a basis γ_1, γ_2 of $R_1 f_*^{an}\mathbb{Z}$ in a punctured neighborhood of $x = 0$, such that on writing $\omega_2 = d/dx\ \omega_1$, the 1-period matrix can be explicited in the following fashion:

$$\begin{pmatrix} \frac{1}{2i\pi}\int_{\gamma_1}\omega_1 & \frac{1}{2i\pi}\int_{\gamma_2}\omega_1 \\ \frac{1}{2i\pi}\int_{\gamma_1}\omega_2 & \frac{1}{2i\pi}\int_{\gamma_2}\omega_2 \end{pmatrix} = \begin{pmatrix} y_1 & y_2 \\ \frac{dy_1}{dx} & \frac{dy_2}{dx} \end{pmatrix}, \text{ where}$$

$$y_1 = \frac{\sqrt{2}}{12}(1-x)^{1/4}\ {}_2F_1\left(\frac{1}{12},\frac{5}{12},1,x\right) \in \bar{\mathbb{Q}}[[x]] ,$$

$$y_2 = y_1\left(\frac{1}{2i\pi}\log\frac{x}{1728} + \text{hol. vanishing at } x = 0\right).$$

Moreover $\tau(J) = \left(\int_{\gamma_2}\omega_1 \Big/ \int_{\gamma_1}\omega_1\right)(1/J)$ is an inverse of the elliptic modular form $J(\tau)$, and τ carries $\mathbb{C}\setminus]-\infty,1]$ into the usual fundamental domain for $SL_2(\mathbb{Z})$.

§ 2. HODGE CYCLES AND PERIOD RELATIONS

2.1 Hodge cycles

Let X denote a smooth projective k-variety. The Hodge-De Rham spectral sequence (1.1.2) defines the Hodge (decreasing) filtration F^p on the abutment $H^{p+q}_{DR}(X)$. Similarly the tensor powers $H^n_{DR}(X)^{\otimes m}$ are endowed with a natural filtration, namely the tensor product filtration

$F^r(H^n_{DR}(X)^{\otimes m}) = \sum_{\Sigma r_j = r} \otimes F^{r_j} H^n_{DR}(X)$; this filtration is compatible

with the projection on the $(n,n,\ldots n)$- component in the Künneth formula (where X^m denotes the m-fold product of X with itself):

$$(2.1.1)\qquad H^{nm}_{DR}(X^m) \simeq \sum_{\Sigma n_j = nm} \bigotimes_{j=1}^{m} H^{n_j}_{DR}(X) .$$

DEFINITION. <u>The Hodge ring of</u> X <u>in degree</u> n, denoted by $Hg^n(X_k)$ or $Hg^n(X)$, is the set of elements of the k-spaces $F^r H^n_{DR}(X)^{\otimes m}$ (for any m,r with $nm = 2r$) whose image under $(2i\pi)^{-r} P_X$ lies in the $\mathbb{Q}$-space $H^n(X^{an}_{\mathbb{C}},\mathbb{Q})^{\otimes m}$; it is a subring of the tensor ring over $H^n_{DR}(X)$. Any element of a $Hg^n(X)$ is called a <u>Hodge cycle</u> (with respect to k).

EXPLANATION. When $k = \mathbb{C}$, one recovers the familiar notion. Indeed, let us consider the Hodge bigrading

$$(2.1.2)\qquad H^n(X^{an},\mathbb{Q})^{\otimes m} \otimes_{\mathbb{Q}} \mathbb{C} = \sum_{p+q=mn} H^{p,q} ,$$

$$(2.1.3)\qquad H^{p,q} = \overline{H^{p,q}} ;$$

then the filtration F^r on H_{DR} induces (through the isomorphism P_X) the filtration $\sum_{p \geq r} H^{p,q}$ on $H^n(X^{an},\mathbb{Q})^{\otimes m} \otimes_{\mathbb{Q}} \mathbb{C}$. Now the condition $t \in Hg^n(X_{\mathbb{C}})$ means that

$(2i\pi)^{-r} P_X(t) \in H^n(X^{an},\mathbb{Q})^{\otimes n} \cap \sum_{p \geq r} H^{p,q}$ for $r = \frac{mn}{2}$; hence

$\overline{P_X(t)} = P_X(t)$ and using (2.1.3) we find:

$$(2.1.4)\qquad t \in Hg^n(X_{\mathbb{C}}) \iff (2i\pi)^{-r} P_X(t) \in H^n(X^{an},\mathbb{Q})^{\otimes m} \cap H^{r,r}$$

$$\iff P_X(t) \in H^n(X^{an},\mathbb{Q})^{\otimes m}(r) \cap H^{0,0} ,$$

for (r) the $r^{\underline{th}}$-tensor product of the (Tate) Hodge structure of type (-1,-1) on $(2i\pi)\mathbb{Q}$.
For an arbitrary algebraically closed subfield k of $\mathbb{C}$, we clearly have $Hg^n(X_k) \subset Hg^n(X_{\mathbb{C}})$. Whether the converse inclusion holds is a very delicate question which we shall not discuss here; the equality would follow from the Hodge conjecture (according to which Hodge cycles should be $\mathbb{Q}$-linear combinations of cohomology classes of subvarieties), or from the less optimistic hope of Deligne, according to which Hodge cycles should be "absolute", see appendix.

2.2 Grothendieck's conjecture

Because any Hodge cycle $t \in Hg^n(X_k)$ belongs to some $H^n_{DR}(X)^{\otimes m}$, the coordinates of $(2i\pi)^{-mn}P_X(t)$ in the basis of $H^n(X^{an}_{\mathbb{C}},\mathbb{Q})^{\otimes m} \otimes_{\mathbb{Q}}\mathbb{C}$ given by $\tilde{\gamma} = \bigotimes_{j=1}^{M} \gamma^*_{l_j}$, $\{l_1,\dots,l_m\} \subset \{1,\dots,\mu\}$,

are homogeneous polynomials, say $q_{t,\tilde{\gamma}}$, of degree m with coefficients in k, in the n-periods $\frac{1}{(2i\pi)^n}\int_{\gamma_j} \omega_i$, $\{i,j\} \subset \{1,\dots,\mu\}$;

here μ denotes $b_n := n^{\underline{th}}$ Betti number of $X_{\mathbb{C}}$.

Because the image of such a Hodge cycle under $(2i\pi)^{-\frac{mn}{2}}P_X$ belongs to $H^n(X^{an}_{\mathbb{C}},\mathbb{Q})^{\otimes m}$, we get a system of equations of the form:

(2.2.1) $q_{t,\tilde{\gamma}}(\text{n-periods}) = (2i\pi)^{-\frac{mn}{2}} H_{\tilde{\gamma}}$, with mn even, and

$\kappa_{\tilde{\gamma}} \in k$.

Moreover if $t \neq 0$, then at least one constant $\kappa_{\tilde{\gamma}}$ in k is non-zero.
This system of equations (for all t and all $\tilde{\gamma}$) defines a closed subscheme Π' of the affine space $\mathbb{A}^{\mu^2}$ over $k'_X = k((2i\pi)^n)$ or at worse $k((2i\pi)^{n/2}))$.

EXAMPLE: polarization. Because X is projective and smooth, its cohomology is polarizable. In particular, there are non-degenerate bilinear forms

(2.2.2) $H^n_{DR}(X) \underset{k}{\otimes} H^n_{DR}(X) \longrightarrow k$

and

(2.2.3) $H^n(X^{an}_{\mathbb{C}},\mathbb{Q}) \otimes_{\mathbb{Q}} H^n(X^{an}_{\mathbb{C}},\mathbb{Q}) \longrightarrow (2i\pi)^{-n}\mathbb{Q}$

which are compatible under P_X on the left side and the double imbedding $k \hookrightarrow \mathbb{C} \hookleftarrow (2i\pi)^{-m}\mathbb{Q}$ on the right side. Taking Poincaré duals, we arrive at a non-zero Hodge cycle $t^n_{pol} \in H^n_{DR}(X)^{\otimes 2}$ (in fact t^n_{pol} lies either in $\Lambda^2 H^n_{DR}$ or in $S^2 H^n_{DR}$ according to the parity of n). There is an associated non-trivial relation (2.2.1), which shows that

(2.2.4) $(2i\pi)^{-n}$ is a homogeneous polynomial of degree 2 in the n-periods with coefficients in k . Hence $(2i\pi)^n \in k(P^n_X)$.

For X = a curve or an Abelian variety, these relations are the classical Riemann relations.
In fact, when n is even, it may happen that not only $(2i\pi)^n$ but also $(2i\pi)^{n/2}$ occurs in (2.2.1), hence belongs to $k(P^n_X)$ (for example such a relation might be provided by the image of $t^{n/2}_{pol}$ in $H^n_{DR}(X)$ under the cup-product).
We now turn to the case $k = \bar{\mathbb{Q}}$.
On eliminating $(2i\pi)$ between the equations (2.2.1) for all t and all $\tilde{\gamma}$, we get a closed subscheme of $\mathbb{A}^{\mu^2}_{\bar{\mathbb{Q}}}$, which is simply the $\bar{\mathbb{Q}}$-Zariski closure of $\prod'$, say $\prod$. When the hypersurface of $\mathbb{A}^{\mu^2}_{\bar{\mathbb{Q}}}$ defined by a polynomial relation $q(\text{n-periods}) = 0$ contains some connected component of this closed subscheme, we say that <u>the relation</u> q <u>comes from Hodge cycles in</u> $Hg^n(X_{\bar{\mathbb{Q}}})$.

REMARK. Since $\pi \notin \bar{\mathbb{Q}}$, it is clear on (2.2.1) that $\prod$ is defined by a homogeneous ideal.

The following statement makes a conjecture of A. Grothendieck precise, see $[42]_{(40)}$,[32] p. 47:

CONJECTURE 1. <u>Every polynomial relation with coefficents in</u> $\bar{\mathbb{Q}}$ <u>in the n-periods comes from Hodge cycles in</u> $Hg^n(X_{\bar{\mathbb{Q}}})$.

As far as I know, there are only two (printed) partial results pointing in the direction of this conjecture:

i) the conjecture is true for $X_{\bar{\mathbb{Q}}}$ = any Abelian variety isogeneous to the power of an elliptic curve with complex multiplication, in accordance with a result of G.V. Chudnovsky.

ii) G. Wüstholz [62] has proved the conjecture in the case of linear relations and for $n = 1$ (this is essentially a result about Abelian varieties, and the only Hodge cycles which are required for linear relations are induced by endomorphisms).

Actually, Grothendieck originally required that the Hodge cycles which occur should be algebraic classes.

2.3 Hodge groups. In this section, which may be omitted at a first reading, we are aiming to shed some light on the "variety of periods" over $k'_X(\Pi')$ or $k(\Pi)$.
Let us consider the following functor: k'_X-algebras $\longrightarrow$ Sets, also denoted by Π' :

(2.3.1) $\Pi'(A) = \{$isomorphisms of A-modules $H^n_{DR}(X) \otimes_k A \longrightarrow H^n(X^{an}_{\mathbb{C}},\mathbb{Q}) \otimes_{\mathbb{Q}} A$ which coincide with P_X on $Hg^n(X_k)\}$.

[This is clearly a sheaf in the fpqc topology on Spec k'_X]. One has $\Pi'(k(P^n_X)) \ni P^n_X$. On the other side, let $G^n_{shg}(X_k)$ denote the subgroup-scheme of $GL(H^n_{DR}(X))$ which fixes the elements of $Hg^n(X_k)$ (read: n-th-special Hodge group of X). The obvious action $G^n_{shg}(X_k) \times \Pi' \longrightarrow \Pi'$ makes Π' formally principal homogeneous, hence Π' is a principal homogeneous space, i.e. representable over k'_X . Representing elements of $\Pi'(A)$ by elements of $GL_\mu(A)$, it is clear that Π' is representable by the affine scheme denoted by the same symbol before. In particular Π' is a smooth k'_X-scheme of dimension $\dim G^n_{shg}(X_k)$. If we want to deal with the Zariski-closure Π of Π' over k , another group comes into play: the "full" Hodge group $G^n_{hg}(X_k)$. Let us first notice that the definition of Hodge cycles extends obviously for negative tensor powers $\otimes m$ ($\otimes{-1}$ = dual), hence on spaces $H^n_{DR}(X)^{\otimes m} \otimes H^n_{DR}(X)^{\otimes -m}$; in such a space, a Hodge cycle is in F^0 and goes into $H^n(X^{an}_{\mathbb{C}},\mathbb{Q})^{\otimes m} \otimes H^n(X^{an}_{\mathbb{C}},\mathbb{Q})^{\otimes -m}$ under P_X . Through the contragredient action on negative tensor powers, $G^n_{shg}(X_k)$ keeps on fixing such Hodge cycles, and denoting by $G^n_{shg}(X_k)$ the "full" sub-group-scheme of $GL(H^n_{DR}(X))$ which fixes them, one finds a split exact sequence (for $n > 0$ such that $H^n_{DR} \neq 0$)

$0 \longrightarrow G^n_{shg}(X_k) \longrightarrow G^n_{hg}(X_k) \longrightarrow \mathbb{G}_m \longrightarrow 0$; the point is that $H^{n\otimes-1}_{DR}$ can be identified (via the polarization) with a twisted $H^n_{DR}(X)(n)$, compatibly with P_X.

As before, we see that the functor of k-algebras, say Π'', which associates to A the set of isomorphisms $H^n_{DR}(X) \otimes_k A \longrightarrow H^n(X^{an}_{\mathbb{C}},\mathbb{Q}) \otimes_{\mathbb{Q}} A$ which coincide with P_X on the Hodge cycles in $(H^n_{DR})^{\otimes m} \otimes (H^n_{DR})^{\otimes -m}$ is representable by a principal homogeneous space under $G^n_{hg}(X_k)$, defined over k, which contains Π. If $\pi \notin k$, we thus have $\dim \Pi = \dim G^n_{shg}(X) + 1$, and the following conditions are equivalent:

(i) The Zariski closure of P^n_X over k is a component of Π,

(ii) the same, with Π'' instead of Π,

(iii) $\deg \operatorname{transc}_k k(P^n_X) = \dim G^n_{shg} + 1$,

(iv) $\deg \operatorname{transc}_{k(\pi)} k(P^n_X) = \dim G^n_{shg}$.

In particular for $k = \bar{\mathbb{Q}}$, (i) is conjecture 1, which is therefore equivalent to any of the following variants (using the transcendence of π):

VARIANT 2. <u>Every polynomial relation with coefficients in $\bar{\mathbb{Q}}$ in the n-periods comes from Hodge cycles in some</u> $H^n_{DR}(X)^{\otimes m} \otimes H^n_{DR}(X)^{\otimes -m}$.

VARIANT 3. <u>One has:</u> $\deg \operatorname{trans}_{\bar{\mathbb{Q}}(\pi)} \bar{\mathbb{Q}}(P^n_X) = \dim G^n_{shg}(X)$.

In the appendix, we shall give some tools for the computation of G^n_{shg}.

EXAMPLE. Here is one of the few non-trivial geometric situation ($\dim X = 2$, $n = 2$) where Grothendieck's conjecture holds trivially true. Let X be a cubic surface in $\mathbb{P}^3_{\bar{\mathbb{Q}}}$. The geometric genus of X is 0, hence the Hodge numbers $h^{2,0}$ and $h^{0,2}$ are 0 (Serre duality). Therefore $H^2(X^{an}_{\mathbb{C}},\mathbb{Z})(1) \simeq \operatorname{Pic} X_{\mathbb{C}}$, which is generated by the classes of 7 among the 27 lines lying on X, see [34] V 4.8. The De Rham cohomology group $H^2_{DR}(X)$ is generated by the cohomology class of theses 7 lines, which are Hodge cycles, thus $G^2_{shg}(X)$ is reduced to the unit element. Conjecture 1 and variant 3 are obviously satisfied, and variant 2 reads in this case: π is transcendental.

2.4 Relative Hodge cycles

Let $f : X \longrightarrow S$ be a projective smooth morphism between smooth (connected) k-variety. The locally free sheaves $\mathbb{R}^{p+q}f_* F^p\Omega^{\cdot}_{X/S}$ on S define a filtration $F^{\cdot}$ on $H^{p+q}_{DR}(X/S)$ (the Hodge filtration), with $Gr^p_F \simeq R^q f_x \Omega^p{}_{X/S}$. One has the associated relative Hodge-De Rham spectral sequence

(2.4.1) $E_1^{pq} = R^q f_* \Omega^p_{X/S} \Rightarrow H^{p+q}_{DR}(X/S)$ (which degenerates at E_1).

Let us consider the tensor product filtration (see 2.1) on the $\mathcal{O}_S$-modules $H^n_{DR}(X/S)^{\otimes m}$. A relative Hodge cycle in $H^n_{DR}(X/S)^{\otimes m}$ is section of $F^r H^n_{DR}(X/S)^{\otimes m}$ over some open dense subset U of S (with $nm = 2r$), whose image under $(2i\pi)^{-r}\, P_{X/S}$ lies in the $\mathbb{Q}$-space of sections of $(R^n f_*^{an}\mathbb{Q})^{\otimes m}$ over $U^{an}_{\mathbb{C}}$.

REMARK. In particular, any relative Hodge cycle is horizonal for the Gauss-Manin connection (acting as a tensor product connection, to the effect that the embedding $H^n_{DR}(X/S)^{\otimes m} \hookrightarrow H^n_{DR}(X^m/S)$ is an embedding of modules with connection). In fact, it follows from the description (2.1.4) together with the global invariant cycle theorem [22] II 4., that a relative Hodge cycle is nothing but a horizontal section of some $H^n_{DR}(X_U/_U)^{\otimes m}$ whose fibre at some (or any) $s \in U$ is a Hodge cycle in the sense of 2.1.

EXAMPLE. Again a polarization on $H^n_{DR}(X/S)$ gives rise to a relative Hodge cycle t^n_{pol}, which shows that $(2i\pi)^n \in$ field of constants in $k(P^n_{X/S})$.

§ 3. PERIOD RELATIONS: THE RELATIVE CASE

3.1 Horizontal cycles. A horizontal cycle in $H^n_{DR}(X/S)^{\otimes m}$ is a section of $H^n_{DR}(X/S)^{\otimes m}$ over some open dense subset U of S, annihilated by the Gauss-Manin connection.

For any such horizontal cycle t, the image $P_{X/S}(t)$ lies $H^0(U^{an}_{\mathbb{C}}, (Rf_*^{an}\mathbb{C})^{\otimes m})$ according to 1.2. Because the map $\pi_1(U^{an}_{\mathbb{C}}, s) \longrightarrow \pi_1(S^{an}_{\mathbb{C}}, s)$ is surjective for any $s \in U(\mathbb{C})$, the stalk of $P_{X/S}(t)$ at s is monodromy-invariant in $H^n(X^{an}_s, \mathbb{C})^{\otimes m}$, i.e. $\pi_1(S^{an}_{\mathbb{C}}, s) \cdot P_{X/S}(t)_s = P_{X/S}(t)_s$.

Now let $\{\theta_j, j=1,\dots,b_n\}$ denote a $\mathbb{C}$-basis of solutions of $H^n_{DR}(X/S)$ in the fraction field of $\mathcal{O}_{S^{an},s}$; e.g. we take any multiple $\alpha^{-1}\int_{\gamma_j}$, $\alpha \in \mathbb{C}^x$. Let us denote by $k(\theta)$ the field generated by $\theta_j(\omega_i)$, $i = j = 1,\ldots,b_n = \mu$, over $k(S)$, and let us set $C :=$ field of constants in $k(\theta)$. Then $\mathrm{Sol}(H^n_{DR}(X/S),k(\theta))$ is the C-vector space spanned by the θ_j ; we denote its dual by E_C , and we write (E,∇) for the C(S)-vector space with connection $H^n_{DR}(X/S) \otimes C(S)$. By integrability of ∇ , there is an isomorphism

$$(3.1.1) \quad T^n_{X/S} : E \otimes_{C(S)} k(\theta) \xrightarrow{\sim} E_C \otimes_C k(\theta) ,$$

$$\text{with } E_C = (E \otimes_{C(S)} k(\theta))^{\nabla} .$$

If $\theta_j = \alpha^{-1}\int_{\gamma_j}$, then $E_C = \alpha\, H^n(X^{an}_s,C)$ and $T^n_{X/S}$ may be identified with $P^n_{X/S}$ after tensoring by the function field of $\mathcal{O}_{S^{an},s}$. On imitating the argument 2.2, one finds that horizontal cycles t in $H^n_{DR}(X/S)^{\otimes m}$ give rise to a collection of equations in the $\theta_j(\omega_i)$'s of the form

$$(3.1.2) \quad q_{t,\tilde{\gamma}}(\theta_j(\omega_i)) = b_{\tilde{\gamma}} \in C .$$

In analogy with conjecture 1 before, one has the following result of Kolchin-Lang-type:

PROPOSITION. <u>The set of relations</u> (3.1.2) <u>given by horizontal cycles generates an ideal of definition of the quantities</u> $\theta_j(\omega_i)$ <u>over</u> C(S) .

□

Let us denote by Θ the closed subscheme of $\mathbb{A}^{\mu^2}_{C(S)}$ defined by the relations (3.1.2); the proposition says that Θ is the C(S)-Zariski closure of the point $(\theta_j(\omega_i)) \in \Theta(k(\theta))$, identified with $T^n_{X/S}$.

EXPLANATION. This proposition belongs properly to differential Galois theory, and it is not the right place to develop it here. We shall only indicate the main steps of the proof, see also [2].

Let us consider the following functor, still denoted by Θ : C(S)-algebras ⟶ Sets,

(3.1.3) $\Theta(A) = \{$isomorphisms of A-modules $\mathcal{E} \otimes_{C(S)} A \xrightarrow{\sim} E_C \otimes_C A$ which coincide with $T^n_{X/S}$ on the horizontal cycles the tensor powers of $\mathcal{E}\}$

An argument similar to 2.3 shows that Θ is representable by the C(S)-scheme denoted before by the same letter. Moreover Θ is a principal homogeneous space under the differential Galois group of $\mathcal{E}$, say $G^n_{diff}(X/S)$, which we define here (ad hoc) to be the closed subscheme of $GL_\mu(\mathcal{E})$ which fixes the horizontal cycles. The proposition then follows easily from the classical description of $G^n_{diff}(X/S)$ (only the case when S is one-dimensional is relevant for the sequel):

(3.1.4) $G^n_{diff}(X/S) = (Aut_{C(S),\nabla}k(\theta)) \times_C C(S)$.

The point here is that all rank-one modules with connection which arise as subquotients of tensor powers of $\mathcal{E}$ are isotrivial, see [2] and II ex.1, VIII ex.6.

The determination of the field C is a rather delicate but important question in the context of G-functions. We shall touch upon it in § 3.4.

REMARK (not used in the sequel). There is an obvious analogy between the proposition just stated and Grothendieck's conjecture. We present here a conjectural link between them. On choosing $\theta_j = \frac{1}{(2i\pi)^n}\int_{\gamma_j}$, we obtain

$$\text{deg transc}_{k(2i\pi)(S)}\, k(P^n_{X/S}) = \text{deg transc}_{k(2i\pi)}\, C + \text{deg transc}_{C(S)}\, k(P^n_{X/S})$$

(3.1.5)

$$= \text{deg transc}_{k(2i\pi)}\, C + \dim G^n_{diff}(X/S) \ .$$

For $k = \bar{\mathbb{Q}}$, the functional analog to Grothendieck's conjecture (variant 3) would identify the left-hand side of (3.1.5) with the dimension of the subgroup $G^n_{shg}(X/S)$ of $GL(H^n_{DR}(X_\eta))$ which fixes the relative Hodge cycles; in fact this would follow from variant 3 itself if $\dim G^n_{shg}(X/S) = \dim G^n_{shg}(X_s)$ for at least some $s \in S(\bar{\mathbb{Q}})$, which is very likely.
On the other hand, when there is no non-zero isotrivial submodule

in $H^n_{DR}(X_{\bar\eta})$, it seems likely that the (possibly transcendental) constants $b_{\tilde\gamma}$ involved in equations (3.1.2) (which generate the field C according to the proposition) are governed by the center of $G^n_{shg}(X/S)$, so that deg $\mathrm{transc}_{k(2i\pi)}C$ should be $\dim Z\, G^n_{shg}(X/S)$ At last, Deligne's hope that Hodge cycles should be absolute (see appendix) would show that $G^n_{shg}(X/S)$ is reductive connected. Putting everything together, we arrive at the following conjectural equality

(3.1.6)? $(G^n_{diff}(X/S))^0 = \mathcal{D}\, G^n_{shg}(X/S)$, which may hold for any k ,

under the assumption that $H^n_{DR}(X/S)$ admits no non-zero constant submodule over any etale covering of S ; here the exponent 0 denotes the connected component, and $\mathcal{D}$ indicates the derived group. See [3] for more information about this conjecture. Anyway, one can show that inclusion $\subseteq$ holds in (3.1.6).

3.2 Mustafin's theorem generalized. We have seen (2.4) that k-linear combinations of relative Hodge cycles are horizontal cycles. Here we present a geometric situation where the converse also holds.

Let S' denote a smooth (not necessarily complete) curve over k , S = the complement of a closed point s_0 in S' , and let $f : X \longrightarrow S$ be a smooth projective morphism.

DEFINITION (Mustafin). We say that f has strong degeneration (shortly S.D.) at s_0 if and only if there exists a smooth k-scheme X' and a projective morphism $f' : X' \longrightarrow S'$ such that:

i) $f'_{|S} = f$

ii) $Y := f'^{-1}(s_0)$ is a union of transversally crossing smooth divisors Y_i entering the fiber with multiplicity one,

iii) the whole cohomology of every stratum

$$Y^{[d]} := \coprod_{i_0<i_1<\ldots<i_d} \bigcap_{j=0}^{d} Y_{i_j} , \quad d = 0,1,\ldots, \quad \text{is}$$

spanned by Hodge cycles.

LEMMA 1 ([50] 3.3). *If two smooth projective morphisms* $f_1 : X_1 \longrightarrow S$, $f_2 : X_2 \longrightarrow S$ *have S.D. at* s_0 , *so does the product* $f_1 \times f_2 : X_1 \times_S X_2 \longrightarrow S$.

□

The following result was proved by G. Mustafin (loc.cit.) in the special case $k = \mathbb{C}$:

THEOREM. Assume that f has strong degeneration at s_0, or that f is an Abelian scheme with multiplicative reduction at s_0 (i.e. the fiber at s_0 of the connected Néron model is a torus). Then every horizontal cycle in $H^n_{DR}(X/S)^{\otimes m}$ is a k-linear combination of relative Hodge cycles.

In fact, Mustafin first shows that if iii) is replaced by the weaker assumption

iv) $H^n(Y^{[0]^{an}}, \mathbb{C})$ is of type $(\frac{n}{2}, \frac{n}{2})$ in the Hodge bigrading,

then the global monodromy invariant cycles in $H^n(X_s^{an}, \mathbb{Q})$ are of type $(\frac{n}{2}, \frac{n}{2})$ (hence if $k = \mathbb{C}$, they are image of relative Hodge cycles in $H^n_{DR}(X/S)$ under $((2i\pi)^{-n/2} P_{X/S})_s$.)

The proof of this fact relies upon a clever use of the Clemens morphism in the theory of the variation of Hodge structure and we shall adapt this argument in order to prove the following lemma in the next section:

LEMMA 2. Let U be a Zariski-dense open subset of S, and let s be any closed point of U. Then under the assumptions i) ii) and iv), the fiber at s of any section t of $H^n_{DR}(X/S)^{\nabla}$ over U is a k-linear combination of Hodge cycles.

On the other hand, Mustafin shows [50] 3.1 that an Abelian scheme may be extended over S' with properties i), ii), iv) if and only if it has multiplicative reduction at s_0 - a condition which is certainly stable under taking products.

Granted lemma 2, the proof of the theorem is achieved on using the embedding $H^n(X/S)^{\otimes m} \hookrightarrow H^n(\underbrace{X \times_S X \times_S \cdots \times_S X}_{\text{m-factors}}/S)$ and applying lemma 1 or the last remark. Note that the theorem can be considered in no way as a corollary of the special case $k = \mathbb{C}$ (at least not directly).

3.3 The limit Hodge structure (survey).

Let us consider a commutative diagram as in 3.2 (dim S = 1)

$$\begin{array}{ccc} X & \hookrightarrow & X' \\ \text{projective } f \ \text{smooth} \downarrow & & \downarrow f' \text{ projective} \\ S & \hookrightarrow & S' \end{array} \qquad \begin{cases} X' \setminus X = Y = f'^{-1}(s_0) \\ S' \setminus S = s_0 \end{cases}$$

satisfying the assumption ii); hence Y is a <u>reduced</u> divisor with simple normal crossings. Let us denote by $\Omega^{\cdot}_{X'/S'}\langle Y\rangle$ the <u>relative De Rham complex with logarithmic poles</u> along Y, defined by the exact sequence

(3.3.1) $$0 \to f'^{-1}\Omega^1_{S'}\langle s_0\rangle \underset{f'^{-1}\mathcal{O}_{S'}}{\otimes} \Omega^{\cdot}_{X'/S'}\langle Y\rangle[-1] \to \Omega^{\cdot}_{X'}\langle Y\rangle \to \Omega^{\cdot}_{X'/S'}\langle Y\rangle \to 0 .$$

This complex has an increasing filtration $W.$ defined by

(3.3.2) $$W_m\Omega^p_{X'/S'}\langle Y\rangle = \Omega^m_{X'/S'}\langle Y\rangle \wedge \Omega^{p-m}_{X'/S'}$$

and a decreasing filtration $F^{\cdot}$, the Hodge filtration. The coherent sheaves of hypercohomology $\mathbb{R}^n f'_*(\Omega^{\cdot}_{X'/S'}\langle Y\rangle)$ are naturally endowed with a connexion ∇' with a logarithmic pole at s_0. The following properties hold:

(3.3.3) a) $(\mathbb{R}^n f'_*(\Omega^{\cdot}_{X'/S'}\langle Y\rangle),\nabla')_{|S} \simeq (\mathbb{R}^n f_*(\Omega^{\cdot}_{X/S}),\nabla)$,

b) the residue of ∇' at s_0 is nilpotent, (Katz)

c) $\mathbb{R}^n f'_*(\Omega^{\cdot}_{X'/S'}\langle Y\rangle)$ is locally free, (Steenbrink)

see [58], [62]. Up to isomorphism, $\mathbb{R}^n f'_*(\Omega^{\cdot}_{X'/S'}\langle Y\rangle)$ is the unique $\mathcal{O}_{S'}$-module with these properties (the <u>canonical extension</u>). The Hodge filtration $\mathbb{R}^n f'_* F^p\Omega^{\cdot}_{X'/S'}\langle Y\rangle$ coincides with the filtration on the abutment of the Hodge-Dr Rham spectral sequence

(3.3.4) $$E_1^{p,q} = R^q f'_*\Omega^p_{X'/S'}\langle Y\rangle \Rightarrow \mathbb{R}^{p+q} f'_*\Omega^{\cdot}_{X'/S'}\langle Y\rangle \quad \text{for} \quad p+q = n .$$

This spectral sequence <u>degenerates at</u> E_1, and the $E_1^{p,q}$ terms as well as the $\mathbb{R}^n f'_* F^p\Omega^{\cdot}_{X'/S'}\langle Y\rangle$ are locally free (Steenbrink). In particular we get a filtration $F^{\cdot}_{DR}$ on

(3.3.5) $$H^n_{DR\ \lim} := \mathbb{H}^n(Y,\Omega^{\cdot}_{X'/S'}\langle Y\rangle \otimes \mathcal{O}_Y)$$
$$\simeq \text{fiber of } \mathbb{R}^n f'_*\Omega^{\cdot}_{X'/S'}\langle Y\rangle \text{ at } s_0 .$$

Now we are looking for an extension of the isomorphism $P^n_{X/S}$ at the point s_0 ; in other words, we seek a natural rational realization of $H^n_{DR\ \lim} \otimes_k \mathbb{C}$.

To this aim, let us extend the scalars to $\mathbb{C}$ and localize the entire situation over a unit disc Δ centered at s_0 :

$$\begin{cases} \Delta \hookrightarrow S'^{an}_{\mathbb{C}} \\ 0 \longmapsto s_0 \end{cases} .$$

We put $\Delta^* = \Delta\setminus\{0\}$, $\mathcal{X} = \mathcal{X}'_\Delta$, $\mathcal{X}^* = \mathcal{X}_{\Delta^*}$. We identify the universal covering $\widetilde{\Delta}^*$ of Δ^* with the upper half plane: $\widetilde{\Delta}^* \xrightarrow{\exp 2i\pi u} \Delta^*$, so that the canonical generator of $\pi_1(\widetilde{\Delta}^*)$ acts on $\widetilde{\Delta}^*$ via $u \longmapsto u+1$. This action induces the local monodromy on the space

(3.3.6) $H^n_{\mathbb{Q}\,\lim} := H^n(\mathcal{X}^* \times_{\Delta^*} \widetilde{\Delta}^*, \mathbb{Q})$.

Because of assumption ii), this local monodromy is unipotent and we denote by N its logarithm divided by $2i\pi$; we have $N^{n+1} = 0$. As wanted, there is a natural isomorphism (Steenbrink, loc.cit.):

(3.3.7) $P^n_{\lim} : H^n_{DR\,\lim} \otimes_k \mathbb{C} \xrightarrow{\sim} H^n_{\mathbb{Q}\,\lim} \otimes_{\mathbb{Q}} \mathbb{C}$.

Each section σ of $\mathcal{X}^* \times_{\Delta^*} \widetilde{\Delta}^* \longrightarrow \widetilde{\Delta}$ defines a family of homotopy equivalences $\varphi_{z,\sigma,m} : \mathcal{X}_z = X^{an}_{s(z)} \longrightarrow \mathcal{X}^* \times_{\Delta^*} \widetilde{\Delta}^*$, $m \in \mathbb{Z}$ arising from the diagram:

(3.3.8)
$$\begin{array}{ccc} \mathcal{X}^* \times_{\Delta^*} \widetilde{\Delta}^* & \xleftarrow{\ \sigma\ } & \widetilde{\Delta}^* \\ \vdots\ \{\varphi_{z,\sigma,m}\}_{m\in\mathbb{Z}} & & \uparrow \ \frac{\log z}{2i\pi} + \mathbb{Z} \\ \mathcal{X}_z & \longrightarrow & \{z\} \in \Delta^* \end{array}$$

which fits into a commutative diagram at the level of cohomology:

(3.3.9)
$$\begin{array}{ccc} H^n_{\mathbb{Q}\,\lim} & \xrightarrow[\sim]{\varphi^*_{z,\sigma,0}} & H^n(\mathcal{X}_z,\mathbb{Q}) \\ \downarrow & & \downarrow \\ H^n_{\mathbb{Q}\,\lim} & \xrightarrow{\varphi^*_{z,\sigma,1}} & H^n(\mathcal{X}_z,\mathbb{Q}) \end{array} .$$

vertical arrows: local monodromy

The weight filtration $M_.$ on $H^{\cdot}_{\mathbb{Q}\,\lim}$ is determined by the two conditions:

a) $N : M_m \longrightarrow M_{m-2}(-1) \quad (= \frac{1}{2i\pi} M_{m-2})$

b) N^m induces an isomorphism

(3.3.10) $\quad Gr^M_{n+m}(H^n_{\mathbb{Q}\,\lim}) \xrightarrow{\sim} Gr^M_{n-m}(H^n_{\mathbb{Q}\,\lim})(-m)$.

Then $(M_\cdot, P^n_{\lim}(F^\cdot_{DR} \otimes \mathbb{C}))$ is a mixed Hodge structure on $H^n_{\mathbb{Q}\,\lim}$, the so-called <u>limit Hodge structure</u> (Schmid [53], Steenbrink [58]), and (3.3.10) is a morphism of Hodge structure.

There exists a morphism of mixed Hodge structure, the <u>Clemens morphism</u> κ_0 , which makes the following diagram commutative:

(3.3.11)
$$\begin{array}{ccc} H^n(Y^{an}_{\mathbb{C}},\mathbb{Q}) & \xrightarrow{\ \kappa_0\ } & H^n_{\mathbb{Q}\,\lim} \\ sp_0^* \uparrow & & \wr\wr \downarrow \varphi^*_{z,\sigma,0} \\ H^n(X'^{an}_{\mathbb{C}},\mathbb{Q}) & \xrightarrow{\ sp_z^*\ } & H^n(X_z,\mathbb{Q}) \end{array}$$

(the symbol "sp" denotes a specialization map).

The <u>local invariant cycle theorem (not used here)</u> states that:

(3.3.12) the image of κ_0 is $\ker N$.

The <u>global invariant cycle theorem</u> states that:

(3.3.13) the two maps sp'^*_z, sp^*_z :

$$\begin{array}{ccc} H^n(X'^{an}_{\mathbb{C}},\mathbb{Q}) & \xrightarrow{\qquad sp'^*_z \qquad} & H^n(X_z,\mathbb{Q}) \\ & \searrow \qquad\qquad \nearrow {\scriptstyle sp^*_z} & \\ & H^n(X^{an}_{\mathbb{C}},\mathbb{Q}) & \end{array}$$

have the same image, namely $H^n(X_z,\mathbb{Q})^{\pi_1(S,s)}$ (Deligne [22] I 4).

This latter result, when combined with (3.3.11), yields:

(3.3.14) the embedding $H^n(X_z,\mathbb{Q})^{\pi_1(S,s)} \xhookrightarrow[(\varphi^*_{z,\sigma,0})^{-1}]{} \ker N$

is a morphism of Hodge structure (Mustafin [50] 2.1.).

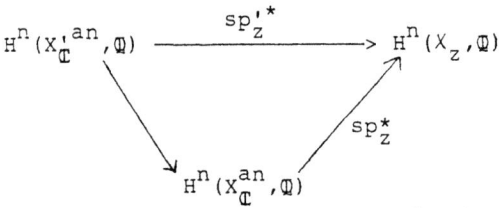

3.4 Proof of lemma 2.

Under the same hypotheses as in 3.3, let us consider the map $a^*_{DR} : H^n_{DR}(X') \longrightarrow H^n_{DR}(Y^{[0]})$ induced by the inclusions of smooth $Y_j \hookrightarrow X'$. Let us consider the morphism $H^n_{DR}(X') \longrightarrow \Gamma(H^n_{DR}(X/S)^{\nabla})$, which arises as the composition of $H^n_{DR}(X') \longrightarrow H^n_{DR}(X)$ and the edge homomorphism $H^n_{DR}(X) \longrightarrow \Gamma(H^n_{DR}(X/S)^{\nabla})$ of the Leray spectral sequence in the category MIC (see II):

$$E_2^{0,n} = (H^0_{DR}(X, R^n f^{DR}_*(\mathcal{O}_X, d)), \text{ trivial connection: } d) \Rightarrow (H^n_{DR}(X), \nabla = d).$$

We first show that $\ker a^*_{DR} \subseteq \ker(H^n_{DR}(X') \to \Gamma(H^n_{DR}(X/S)^{\nabla})$. Via the isomorphisms, $P_{Y^{[0]}}$, P_X, $(P_{X/S})_z$ respectively, this is equivalent to showing that the kernel of the specialization map

$sp^*_z : H^n(X'^{an}_{\mathbb{C}}, \mathbb{Q}) \longrightarrow H^n(X_z, \mathbb{Q})^{\pi_1(S,s)}$ contains the kernel of the map

$$a^*_{\mathbb{Q}} : H^n(X'^{an}_{\mathbb{C}}, \mathbb{Q}) \longrightarrow H^n(Y^{[0]\,an}_{\mathbb{C}}, \mathbb{Q}) .$$

Because the inclusions $Y_j \hookrightarrow X'$ which induce $a^*_{\mathbb{Q}}$ factorize through Y, $a^*_{\mathbb{Q}}$ factorizes through $H^n(Y^{an}_{\mathbb{C}}, \mathbb{Q})$, and even through $Gr^W_n H^n(Y^{an}_{\mathbb{C}}, \mathbb{Q})$ because it is a morphism of mixed Hodge structures. Recall that the weight filtration on $H^n(Y^{an}_{\mathbb{C}}, \mathbb{Q})$ is defined by the spectral sequence

(3.4.2) $E_1^{p,q} = H^q(Y^{[p]an}_{\mathbb{C}}, \mathbb{Q}) \Rightarrow H^{p+q}(Y^{an}_{\mathbb{C}}, \mathbb{Q})$,

whose term E_2 is canonically isomorphic to $H(\Sigma_Y, \underline{H}(Y^{[\ ]an}_{\mathbb{C}}))$:

(3.4.3) $E_2^{p,q} = H^p(\Sigma_Y, \underline{H}^q(Y^{[\ ]an}_{\mathbb{C}}))$,

where Σ_Y denotes the simplicial complex of intersections of the components of Y, and $\underline{H}^q(Y^{[\ ]an}_{\mathbb{C}})$ the local system on Σ_Y defined by $\sigma = \{i_0 < \ldots < i_d\} \longmapsto H^q(\bigcap_{j=0}^{d} Y^{an}_{i_j,\mathbb{C}}, \mathbb{Q})$. The spectral sequence (3.4.2) degenerates at E_2, and the natural isomorphism

(3.4.4) $Gr^W_n H^n(Y^{an}_{\mathbb{C}}, \mathbb{Q}) = E_2^{0,n}$

composed with the natural embedding

(3.4.5) $E_2^{0,n} \hookrightarrow H^n(Y^{[0]an}_{\mathbb{C}}, \mathbb{Q}) = \coprod H^n(Y^{an}_{j,\mathbb{C}}, \mathbb{Q})$

(which arises from the easy remark that elements of $H^0(\Sigma_Y, \underline{H}^n(Y_{\mathbb{C}}^{[\ ]an}))$ are nothing but "compatible" cohomology classes on the components $Y^{an}_{j,\mathbb{C}}$)
is the quotient map of $a^*_{\mathbb{Q}}$ through $Gr^W_n H^n(Y^{an}_{\mathbb{C}}, \mathbb{Q})$, which turns out thereby to be injective.
Because the Clemens morphism κ_0 is a morphism of mixed Hodge structure, the square (3.3.11) gives rise to a commutative diagram of morphisms of Hodge structure (using 3.3.14):

$$
\begin{array}{ccc}
H^n(Y^{[0]an}_{\mathbb{C}}, \mathbb{Q}) & & \\
\uparrow\cup & & \\
Gr^W_n H^n(Y^{an}_{\mathbb{C}}, \mathbb{Q}) & \xrightarrow{Gr_n \kappa_0} & Gr^M_n(\mathrm{Ker}\, N) \\
\uparrow & & \uparrow Gr_n(\varphi_{z,\sigma,0})^{-1} \\
Gr^W_n H^n(X'^{an}_{\mathbb{C}}, \mathbb{Q}) & \xrightarrow{sp^*_z} & H^n(H_z, \mathbb{Q})^{\pi_1(S,s)}
\end{array}
\qquad (3.4.6)
$$

Since $(\varphi_{z,\sigma,0})^{-1}$ is injective and $W_{n-1}H^n(X_z,\mathbb{Q}) = 0$, $Gr_n(\varphi_{z,\sigma,0})^{-1}$ remains injective.
This is enough to prove the desired inclusion of kernels, by diagram-chasing. Moreover the induced diagram

$$
\begin{array}{ccc}
H^n(Y^{[0]an}_{\mathbb{C}}, \mathbb{Q}) & & \\
\uparrow\cup & & \\
H^n(X'^{an}_{\mathbb{C}}, \mathbb{Q})/_{\ker a^*_{\mathbb{Q}}} & \twoheadrightarrow & H^n(X_z, \mathbb{Q})^{\pi_1(S,s)}
\end{array}
\qquad (3.4.7)
$$

exhibits $H^n(X_z,\mathbb{Q})^{\pi_1(S,s)}$ as a subquotient Hodge structure of $H^n(Y^{[0]an}_{\mathbb{C}}, \mathbb{Q})$ (1) (the surjectivity of the arrow at the bottom follows from the global invariant cycle theorem (3.3.13)).

In order to prove lemma 2, we may assume that $U = S$ without loss of generality. Let us now use the hypothesis

---

(1) this simplifies Mustafin's original argument which ought to use the local invariant cycle theorem in addition.

$H^n(Y_{\mathbb{C}}^{[0]an},\mathbb{Q}) = H^{n/2,n/2}$. By (3.4.7), it follows that $H^n(X_z,\mathbb{Q})^{\pi_1(S,s)}$ is a Hodge structure of type $(n/2,n/2)$. Looking at the twin diagram

$$(3.4.8)\qquad \begin{array}{ccc} H^n_{DR}(Y^{[0]}) & & \\ \uparrow & & \\ H^n_{DR}(X')/_{\ker a^*_{DR}} & \twoheadrightarrow & (\Gamma(H^n_{DR}(X/S)^\nabla)) \end{array}$$

we find that $F^{n/2}_{DR}\Gamma(H^n_{DR}(X/S)^\nabla) = \Gamma(H^n_{DR}(X/S))^\nabla$,

and that $\frac{1}{(2i\pi)^{n/2}} P_{X/S}(\Gamma(H^n_{DR}(X/S)^\nabla)) \subseteq (\Gamma R^n f_*^{an}\mathbb{Q}) \underset{\mathbb{Q}}{\otimes} k$

because $\frac{1}{(2i\pi)^{n/2}} P_{Y^{[0]}}(H^n_{DR}(Y^{[0]})) \subseteq H^n(Y_{\mathbb{C}}^{[0]an},\mathbb{Q}) \underset{\mathbb{Q}}{\otimes} k$.

This means that horizontal cycles in $H^n_{DR}(X/S)$ are k-linear combinations of relative Hodge cycles, as required.

□

Hence theorem 3.2 is now proved.

□

REMARK 1. Some hypothesis on f is necessary in order to get the conclusion of lemma 2. However, at least two other methods lead to results of the same kind.
Firstly, there is a method based on the use of "local Torelli" theorems, see [50] 1, which shows that the conclusion of lemma 2 holds for any non-isotrivial family of 4-dimensional cubic hypersurfaces. Secondly, there is a method based on the fact that connected components of monodromy groups are normal subgroups of generic "special Mumford-Tate groups" (see appendix thereafter and [3]), which shows for instance that any non-isotrivial simple Abelian scheme of odd relative dimension and not of type IV satisfies the conclusion of lemma 2, cf. next chapter. In these two examples, one can even replace Hodge cycles by algebraic cycles.

REMARK 2. In complete analogy with 2.2, the relative Hodge cycles elements of $Hg^n(X/S)$ give relations of type (2.2.1) with coefficients in $k'_{X/S} = k(S)((2i\pi)^n)$ or $k(S)((2i\pi)^{n/2})$ between the relative n-periods. These relations define a closed subscheme Π' of $\mathbb{A}^{\mu^2}_{k'_{X/S}}$ (with μ = dimension of the fibers of $H^n_{DR}(X/S)$), whose k(S)-closure we denote by Π.

Let us consider the C(S)-scheme Θ of solutions $\theta_j = \frac{1}{(2i\pi)^n}\int_{\gamma_j}$ introduced before. Theorem 3.2 reads: $\Theta = \Pi' \times_{\operatorname{Spec} k'_{X/S}} \operatorname{Spec} C(S)$.

By proposition 3.1, we know that:

Θ is the C(S)-Zariski closure of its $k(P^n_{X/S})$-point $P^n_{X/S} = T^n_{X/S}$.

We deduce immediately:

Π is the k(S)-Zariski closure of $P^n_{X/S}$ in $\mathbb{A}^{\mu^2}_{k(S)}$. in other words (under the hypothesis of lemma 2):

COROLLARY. <u>The set of relations given by elements of</u> $Hg^n(X/S)$ <u>(after eliminating the factors</u> $2i\pi$) <u>generates the ideal of definition of the relative</u> n-<u>periods over</u> $k(S)$.

In particular the field of constants C is $k'_{X/S} \cap \mathbb{C} = k((2i\pi)^n)$ or $k((2i\pi)^{n/2})$. Moreover, if we had chosen $\theta_j = \frac{1}{(2i\pi)^{n/2}}\int_{\gamma_j}$ rather than the period-solutions, the field of constants would have been k .

§ 4. <u>PERIODS AND G-FUNCTIONS</u>

4.1 In this paragraph, we show essentially, under some hypotheses of the type "strong degeneration at s_0 ", that the Taylor expansions of certain locally (monodromy-) invariant relative periods, with respect to a local parameter x centered at s_0 , are G-functions. By the theorem proved in chapter V, appendix, it suffices to show that the Taylor expansions belong to $\bar{\mathbb{Q}}[[x]] \subseteq \mathbb{C}((x))$, because relative periods are at any rate solutions of geometric differential equations. Here are the precise statements.

4.2 Let S' be a smooth connected curve (not necessarily complete) over $k = \bar{\mathbb{Q}}$, and let S be the complement of a closed point s_0 in S' . Let x denote a local parameter of S' at s_0 . Let $f : X \longrightarrow S$ be a proper smooth morphism, and let us put $\nu = \dim X = n+1$.

THEOREM 1. <u>Assume in addition that</u> f <u>is an Abelian scheme, with multiplicative reduction at</u> s_0 (see 3.2). <u>Then there exists a</u>

basis of sections $\omega_1,\dots,\omega_{2n}$ of $H^1_{DR}(X/S)$ over some dense open subset of S, such that the Taylor expansions in x of the n locally invariant relative periods $\frac{1}{2i\pi}\int_{\gamma_j}\omega_i$ are G-functions (where $\{\gamma_1,\dots,\gamma_n\}$ = maximal set of independent 1-cycles locally invariant around s_0).

Before giving theorem 2, we need some further

NOTATIONS. Let $h^n(\Sigma_Y)$ denote the dimension of the nth group of cohomology of the dual graph of intersections of Y, as before. Let $2i\pi N^*$ denote the logarithm of the local monodromy around s_0, acting on the sheaf $\mathbb{R}_n(f^{an}_{\mathbb{C}})_*(\mathbb{Q})$ localized over the inclusion of a small punctured disk Δ^* centered at s_0.

In accordance with the formalism of the weight filtration, we denote by $M_0\mathbb{R}_n(f^{an}_{\mathbb{C}})_*(\mathbb{Q})_{|\Delta^*}$ the image of $(2i\pi N^*)^n$. We call M_0-n-period any relative n-period over a cycle in $M_0\mathbb{R}_n(f^{an}_{\mathbb{C}})_*(\mathbb{Q})_{|\Delta^*}$.

THEOREM 2. Assume that f extends to a projective morphism $f' : X' \longrightarrow S'$, with X' regular, and with $Y := f'^{-1}(s_0)$ = a union of smooth divisors Y_i with normal crossings entering the fibre with multiplicity 1. Then there exists a basis of sections ω_i of $H^n_{DR}(X/S)$ over some dense open subset of S, such that for any section γ of $M_0\mathbb{R}_n(f^{an}_{\mathbb{C}})_*(\mathbb{Q})_{|\Delta^*}$, the Taylor expansion in x of the relative M_0-n-periods $\frac{1}{(2i\pi)^n}\int_\gamma \omega_i$ are globally bounded G-functions. Moreover $\dim(M_0\mathbb{R}_n(f^{an}_{\mathbb{C}})_*(\mathbb{Q}))_z = h^n(\Sigma_Y)$, for any $z\in\Delta^*$.

More precisely, we shall show under the hypotheses of theorem 2 that the Taylor expansion of $\frac{1}{(2i\pi)^{\nu-1}}\int_\gamma \omega_i$ is the diagonal of an algebraic function in $k[[\underline{x}]] = k[[x_1,\dots,x_\nu]]$, hence the diagonal of a rational function in 2ν variables according to a result of Denef-Lipshitz ("algebraic power series and diagonals", Jour. of number theory 26 (1987) 46-67) (this gives a positive answer to Christol's conjecture stated in I.4 in this particular case, for n = 1). In fact theorem 1 might be deduced in principle from theorem 2 (n = 1) via the theory of relative Picard functors, but we shall not follow this way and shall rather show how different methods lead to similar results: the method of Néron models for theorem 1, and the theory of the limit Hodge structure for theorem 2.

Anyway, we shall carry most of the proofs without assuming $k = \bar{\mathbb{Q}}$.

4.3 Proof of theorem 1.

We denote by X' the connected Néron model of X_η over S' , so that $X'|_S = X$ (this is a slight distorsion of our previous notations 3.3). Working locally over the image of a unit disk Δ in $S'_{\mathbb{C}}$ centered at s_0 , we also write $X = X'_\Delta$, $X^* = X_{\Delta^*}$, etc.... Let us consider the "exponential" exact sequence of sheaves on Δ :

$$(4.3.1) \qquad 0 \longrightarrow \Gamma \longrightarrow \underline{\mathrm{Lie}}\ X/_\Delta \longrightarrow X \longrightarrow 0$$

with $\Gamma = R_1(f_\Delta)_*(\mathbb{Z})$.

The restriction $\Gamma|_{\Delta^*}$ is a lattice bundle in $\underline{\mathrm{Lie}}\ X^*/_{\Delta^*}$. Let us denote by Γ^f the disjoint union of the sections of Γ passing through the fiber of Γ at 0. As the fibre X_0 of X at 0 is semi-stable (a torus, in our case), sections of $\underline{\mathrm{Lie}}\ X/_\Delta$ are linear combinations of sections Γ^f with coefficients in $\mathcal{O}_\Delta$. This implies that $\underline{\mathrm{Lie}}\ X/_\Delta$ is a quotient bundle of the canonical extension ([21] 5.1) of $\mathrm{Hom}(H^1_{DR}(X^*/_{\Delta^*}),\mathcal{O}_{\Delta^*}) \simeq \Gamma \otimes \mathcal{O}_{\Delta^*}$. On denoting the neutral section of f' by e , we obtain by duality that $e^*\Omega^1_{X/\Delta}$ is a subbundle of the canonical extension $\mathcal{E}_{can}$ of $H^1_{DR}(X^*/_{\Delta^*})$, (in fact $e^*\Omega^1_{X/\Delta}$ is the Hodge subbundle F^1_{can} , but we won't need this fact; caution: $\mathcal{E}_{can}$ is <u>not</u> $H^1_{DR}(X/\Delta)$, the latter $\mathcal{O}_\Delta$-module being not coherent).

Let us prove that $\Gamma^f_z \otimes \mathbb{Q}$ is the set of $\pi_1(\Delta^*,z)$ - invariant cycles $H_1(X_z,\mathbb{Q})$ for any $z \in \Delta^*$, that us to say

$$(4.3.2) \qquad H^0(\Delta,\Gamma \otimes \mathbb{Q}) = H^0(\Delta^*,\Gamma_{|\Delta^*} \otimes \mathbb{Q}) .$$

The m-torsion points of X_z form a local system over Δ^* which may be identified with $\Gamma/m\Gamma$ via (4.3.1). Elements of the maximal constant subsystem $H^0(\Delta^*,\Gamma_{|\Delta^*}/m\Gamma_{|\Delta^*})$ are m-torsion sections of $X^*/_{\Delta^*}$, thus they extend to the Néron model. Because $X/_\Delta$ is the neutral component of the Néron model, they extend to sections of X/Δ after multiplying by some $M \in \mathbb{N}^{\times}$ (independent of m). Taking inverse limit over $m = l^n$, $n \to \infty$, l = a prime not dividing M , we get

$$H^0(\Delta, \mathbb{Z}_1 \otimes \Gamma) = H^0(\Delta^*, \mathbb{Z}_1 \otimes \Gamma_{\Delta^*}) \text{ , whence (4.3.2).}$$

We now use the fact that X_0 is a torus: it follows that $\dim H^0(\Delta, \Gamma \otimes \mathbb{Q}) = \dim \Gamma_0 \otimes \mathbb{Q} = n$.
Let us consider the periods with respect to a basis of sections of $e^*\Omega^1_{X'/S'}$ over a Zariski neighborhood of s_0 , and to a basis of $H^0(\Delta, \Gamma \otimes \mathbb{Q})$. We denote these locally invariant periods by

$$(4.3.3) \qquad \frac{1}{2i\pi} \int_{\gamma_j} \omega_i \qquad i = 1,\ldots,n \ , \quad j = 1,\ldots,n \ .$$

At 0, these relative periods specialize to the periods of the $\omega_i(0) \in H^0(X'_{s_0}, \Omega^1)$ on the torus X'_{s_0} ; but these periods are nothing but residues, which belong to the ground field k . More precisely, the pairing

$$H^0(X'_{s_0}, \Omega^1) \otimes H_1(X_0, \mathbb{Q}) \xrightarrow{\frac{1}{2i\pi}\int = \mathrm{Res}} k$$

is non-degenerate, being induced by the isomorphism $P^1_{X_{s_0}}$. From this, it follows clearly that the <u>evaluation</u> at 0 of the upper square of the matrix

$$(4.3.4) \qquad \left(Y = \frac{1}{2i\pi}\int_{\gamma_j} \omega_i \right) \quad \begin{matrix} i = 1,\ldots,2n \\ j = 1,\ldots,n \end{matrix}$$

(where $\{\omega_i\}_{i=1,\ldots,2n}$ = basis of sections of E_{can} over $U \ni s_0$, $\{\omega_1,\ldots,\omega_n\}$ = the basis already chosen for the subbundle $e^*\Omega^1_{X'/S'}$) is an <u>invertible</u> matrix in $M_n(k)$. By Nakayama's lemma, we can moreover choose $\omega_{n+1},\ldots,\omega_{2n}$ so that the lower rectangle of $Y(0)$ vanishes.
Now let $G \in M_{2n}(k\{x\})$ represent $\nabla(xd/dx)$ in this basis (recall that $k\{x\}$ denotes the henselization of $k[x]$ at the ideal (x)). Because E_{can} is the canonical extension, $G(0)$ is nilpotent, hence normalized in the sense of III 1. Thus Y can be written in the form

$$(4.3.5) \qquad Y = Y_G D \text{ , with } Y_G = \text{normalized uniform part of the solution and } D \in M_{2n,n}(k) \ .$$

Looking at formula III (13), we see that all entries of the $m^{\underline{th}}$ coefficient Y_m of Y are linear combinations with coefficients in k of the entries of $Y(0)$, hence belong to k, and we get an expansion $Y \in M_{2n,n}(k[[x]])$. On the other hand, it is not difficult to see that the components of Y satisfy differential equations which come from geometry (use for instance the degenerate Leray spectral sequence $R^0 x_*^{DR}(R^n f_{U*}^{DR}(\mathcal{O}_X,d),\nabla) \Rightarrow (R^n(f\circ x)_*(\mathcal{O}_X,d),\nabla))$. Thus according to the theorem stated in ch. V, appendix, the expansion of the components of Y (in the variable x) are G-functions for $k = \bar{\mathbb{Q}}$.

□

4.4 Proof of theorem 2.

This proof has two parts: a first step consists in constructing locally invariant periods by means of diagonal of algebraic functions (which are globally bounded G-functions for $k = \bar{\mathbb{Q}}$, see I 3.2 and I 4.2), building upon ideas of G. Christol [16] 5. The second step consists in showing that the periods just constructed exhaust the (locally invariant) M_0-n-periods. We follow here the notations of 3.3.

The assumption that the components of $Y = f'^{-1}(s_0)$ have simple normal crossings may be restated as follows: (with $\nu = n+1$)

(4.4.1) there exists a covering of X' be affine open subsets $U_1, U_2, \ldots$ such that U_l admits coordinates $x_1,\ldots,x_\nu$ in terms of which Y_i is defined by the equation $x_i = 0$ (or by the equation $1 = 0$, if Y_i does not meet U_l), and in terms of which the uniformizing parameter x at s_0 is $x = \prod_{m=1}^{\nu} x_m^{e_m}$, with $e_m = 0$ or 1.

Let us assume that $Y^{[n]} \neq \phi$, there exist $\nu = n+1$ components of Y, say $Y_1,\ldots,Y_\nu$, such that $\bigcap_{i=1}^{\nu} Y_i$ contains a point $P \in X'$. Let us consider an element U^P of the covering which contains P; hence for this particular open subsets, all e_m are 1, i.e. $x = x_1 x_2 \cdot\ldots\cdot x_\nu$. In other words the fibre U^P_η of U^P over the generic point η of S' is étale over a familiar scheme W introduced in I 3.3 (after extension of the scalars from $k(x)$ to $k(S)$); explicitly there is an étale map

$$(4.4.2) \qquad U^P_\eta \xrightarrow{\pi} W = \operatorname{Spec} k(S)[x_1,\ldots,x_\nu]/_{(x-x_1x_2,\ldots,x_\nu)} .$$

Because π is étale, the Leray spectral sequence in De Rham cohomology reduces to an isomorphism

$$(4.4.3)\qquad E_2^{\nu-1,0} = H_{DR}^{\nu-1}(W,\pi_*\mathcal{O}_{U_\eta^p}) \simeq H_{DR}^{\nu-1}(U_\eta^p) .$$

As in I 3.3, we associate to the tautological horizontal map

$$(4.4.4)\qquad H_{DR}^0(U_\eta^p/W) = \pi_*\mathcal{O}_{U_\eta^p} \xrightarrow{\theta_p} k((x_1,\ldots,x_\nu))$$

another horizontal map

$$(4.4.5)\qquad H_{DR}^{\nu-1}(W,\pi_*\mathcal{O}_{U_\eta^p}) \xrightarrow{\Delta_{\nu,\theta_p}} k((x)) .$$

In more down-to-earth terms, any $\omega \in H_{DR}^{\nu-1}(U_\eta^p) \simeq \Gamma\Omega^{\nu-1}_{U_\eta^p}/d\Omega^{\nu-1}_{U_\eta^p}$ is represented in a unique way by a differential form $\omega = h^p \dfrac{dx_2\cdot\ldots\cdot dx_\nu}{x_2\cdot\ldots\cdot x_\nu}$, where $h^p \in k((\underline{x}))$ is algebraic over $k(\underline{x})$, and the value at ω of the inverse of (4.4.3) composed with Δ_{ν,θ_p} is the diagonal $\Delta_\nu(h^p) \in k((x))$.

On the other hand, the inclusion $i_p : U^p \hookrightarrow X'$ gives rise to a third horizontal map

$$(4.4.6)\qquad H_{DR}^{\nu-1}(X_\eta) \longrightarrow H_{DR}^{\nu-1}(U_\eta^p) .$$

Composing the inverse of (4.4.2) with (4.4.5) and (4.4.6), we get a horizontal map:

$$(4.4.7)\qquad H_{DR}^{\nu-1}(X_\eta) \longrightarrow k((x)) .$$

Because of the property 3.3.3 b) of the canonical extension $H_{DR}^{\nu-1}(X'/S'\langle Y\rangle)$, the map (4.4.7) induces a horizontal map:

$$(4.4.8)\qquad T_p : H_{DR}^{\nu-1}(X'/S'\langle Y\rangle) \longrightarrow k[[x]] .$$

This map admits the following analytic description. Let ω be a section of $H_{DR}^{\nu-1}(X'/S'\langle Y\rangle)$ over a Zariski neighborhood of s_0. Write $i_{\mathfrak{p}}^*\omega = h^p \dfrac{dx_2\cdot\ldots\cdot dx_\nu}{x_2\cdot\ldots\cdot x_\nu}$ with $h^p \in k[[\underline{x}]]$. Let us next localize the situation over the image of a unit disk Δ centered at s_0, and let us consider the "vanishing cycle" $\gamma_{p,z}$

defined by $|x_2| = \dots = |x_\nu| = \varepsilon$ (small) and $x_1 x_2 \dots x_\nu = x(z)$ in $H_{\nu-1}(U^P_z, \mathbb{Z})$. These cycles glue together to give a cycle $i_{P*}\gamma_P \in H^0(\Delta, R^{\nu-1}(f_\Delta)_*\mathbb{Q})$, and we have the formula

$$(4.4.9) \qquad T_P(\omega) = \Delta_\nu(h^P) = \frac{1}{(2i\pi)^{\nu-1}} \int_{i_{P*}\gamma_P} \omega ,$$

see 1.1 Ex.2.

This formula shows that $T_P(\omega)$ is a relative period of f which extends holomorphically at 0. It is clear that $(i_{P*}\gamma_P)_z$ is invariant under the action of $\pi_1(\Delta^*, z)$ on $H_{\nu-1}(X_z, \mathbb{Q})$ for any $z \in \Delta^*$, and we claim that

(*) <u>the $\mathbb{Q}$-span of these cycles</u> (for various P, if any) <u>is</u> $\operatorname{Im}(2i\pi N_z^*)^{\nu-1}$, where $2i\pi N_z^* = -2i\pi\, {}^tN_z$ denotes the logarithm of the monodromy acting on $H_{\nu-1}(X_z, \mathbb{Q}) = H_n(X_z, \mathbb{Q})$. By duality, this means that the orthogonal space is $\operatorname{Ker} N_z^n$. Via the map $\varphi^*_{z,\sigma,0}$ of (3.3.9), (*) also means that the orthogonal of the prolongation of these cycles at $z=0$ is precisely $\operatorname{Ker} N^n = M_{2n-1} H^n_{\mathbb{Q}\,\lim}$.

In order to prove this, one may extend the scalars from $\mathbb{Q}$ and $\mathbb{k}$ to $\mathbb{C}$. Note that the evaluation $T_P(0)$ of the map (4.4.8) at $z=0$ is a map $H^n_{DR\,\lim} \longrightarrow \mathbb{C}$, and all we have to show is that

$$(4.4.10) \qquad \mathbb{C}\langle T_P(0)\rangle_P^{\perp} = M_{2n-1} H^n_{DR\,\lim}, \text{ and}$$

$$(4.4.11) \qquad \operatorname{Codim} M_{2n-1} H^n_{DR\,\lim} = h^n(\Sigma_Y) .$$

Let ω be a section of $H^n_{DR}(X'/S'\langle Y\rangle)$ over a Zariski neighborhood of s_0, represented by $i_P^*\omega = h^P \dfrac{dx_2 \dots dx_{n+1}}{x_2 \dots x_{n+1}}$; hence the fibre (denoted $\omega|_0$) of ω at s_0 belongs to $H^n_{DR\,\lim}$. Moreover by (4.4.9), we have

$$T_P(0) \cdot \omega|_0 = \Delta_\nu(h^P)(0) = h^P(\underline{0}) = \operatorname{Res}_P(i_P^*\omega \wedge f^*dx/x) .$$

Therefore $\omega|_0 \in \mathbb{C}\langle T_P(0)\rangle_P^{\perp} \iff \forall P \in Y^{[n]}$, the class of $i_P^*\omega \wedge f^*dx/x$

in $Gr_n \Omega^{n+1}_{X'}\langle Y\rangle(U^P)$ vanishes,

$\iff \sum_P$ (Poincaré) $\operatorname{Res}_P(i_P^*\omega \wedge f^*dx/x) = 0$.

To go further, we shall use Steenbrink's double complex of sheaves on $Y^{an} = X_0$:

(4.4.12) $A^{p,q} = \Omega_X^{p+q+1}\langle Y^{an}\rangle / W_q\Omega_X^{p+q+1}\langle Y^{an}\rangle$, with differentials

$$\begin{cases} d' & \text{induced by the exterior differential,} \\ d'' & \text{induced by the cup product } \wedge f^* dx/x \,. \end{cases}$$

One defines an increasing filtration W. on the associated simple complex $A^{\cdot}$ in the following way:

(4.4.13) $W_r A^{p,q} = W_{2q+r+1}\Omega_X^{p+q+1}\langle Y^{an}\rangle / W_q\Omega_X^{p+q+1}\langle Y^{an}\rangle$

so that

(4.4.14) $\mathrm{Gr}_r^W A^{\cdot} = \bigoplus_q \mathrm{Gr}_r^W A^{\cdot,q}[-q]$

$= \bigoplus_{j \geq |r|} a_*\Omega^{\cdot}_{Y^{[j]an}}[-j]$ by taking Poincaré residues;

here "a" stands for the projection of any stratum $Y^{[j]an} \longrightarrow Y^{an}$ (take care that Steenbrink's $\tilde{Y}^{[q+1]}$ is our $Y^{[q]an}$). In particular, one finds $A^{\cdot} = W_n A^{\cdot}$, and

(4.4.15) $$\begin{cases} \mathrm{Gr}_n^W A^{\cdot} \simeq \mathrm{Gr}_{n+1}^W \Omega_X^{n+1}\langle Y^{an}\rangle[-n] \,, \\ \mathbb{H}^n(\mathrm{Gr}_n^W A^{\cdot}) \simeq H^0(Y^{[n]}) \,. \end{cases}$$

Let us now record the fundamental results about $(A^{\cdot}, W.)$ (see [58] 4.16 and 5.10):

(4.4.16) The map $\tilde{\theta} : \Omega^{\cdot}_{X/\Delta}\langle Y^{an}\rangle \otimes_{\mathcal{O}_X} \mathcal{O}_{Y^{an}} \longrightarrow A^{\cdot}$ given by $\wedge(-1)^p f^* dx/x$ in degree p is a quasi-isomorphism :

$$H^n_{DR\ \lim} \xrightarrow[\approx]{\tilde{\theta}} \mathbb{H}^n(A^{\cdot}) \,,$$

(4.4.17) The image under $\tilde{\theta} \circ P^{-1}_{\lim}$ of the monodromy filtration M. coincides with the filtration induced by W. on $\mathbb{H}^n(A^{\cdot})$. After these preliminaries, let us return to our question (4.4.10). By using the Čech bicomplex $C^{\cdot}(\{U^p \cap Y\}, \Omega^{\cdot}_{X'/S'}\langle Y\rangle \otimes_{\mathcal{O}_{X'}} \mathcal{O}_Y)$ in order to compute $H^n_{DR\ \lim}$, one notices that the map

(4.4.18) $\mathbb{H}^n(A^\cdot) \longrightarrow \mathbb{H}^n(Gr_n^W A^\cdot) \simeq H^0(Y^{[n]})$, induced by

$A^\cdot = W_n A^\cdot \longrightarrow Gr_n^W A^\cdot$, coincides (after composition with $\tilde{\theta}$)

with the map $H^n_{DR\ \lim} \longrightarrow H^0(Y^{[n]})$ defined by

$\omega_{|0} \longmapsto \sum_p$ (Poincaré) $Res_p(i_p^* \omega \wedge f^* dx/x)$, up to a sign $(-1)^n$. On the other hand, the "weight" spectral sequence

(4.4.19) $E_1^{-n,2n} = \mathbb{H}^n(Gr_n^W A^\cdot) \Rightarrow \mathbb{H}^n(A^\cdot)$

degenerates at E_2 [58]4.20, and since $E_1^{-n-1,2n} = 0$, it gives rise to a commutative diagram

(4.4.20)
$$\begin{array}{ccc} \mathbb{H}^n(A^\cdot) & \dashrightarrow & \\ \downarrow & & \\ Gr_{2n}^M \mathbb{H}^n(A^\cdot) = E_\infty^{-n,2n} = E_2^{-n,2n} = \mathrm{Ker}\ d_1 & \subset & E_1^{-n,2n} \end{array}$$

It is a purely formal fact about (degenerate) spectral sequences that the dotted arrow is the map (4.4.18). It results from all this discussion that $\omega_{|0} \in \mathbb{C}\langle T_p(0)\rangle_p^\perp \Longleftrightarrow$ the class of $\omega_{|0}$ in $Gr_{2n}^M H^n_{DR\ \lim}$ vanishes. This establishes our aim (4.4.10).

Now for (4.4.11): we have codim $M_{2n-1} H^n_{DR\ \lim} = \dim Gr_{2n}^M = \dim \mathrm{Ker}\ d_1$ by (4.4.20). But it happens that $d_1 : H^0(Y^{[n]an}) \longrightarrow H^2(Y^{[n-1]an})$ is dual to the combinatorial coboundary $\delta_1 : H^0(Y^{[n-1]an}) \to H^0(Y^{[n]an})$, up to some Tate twist, see [58] bottom of p. 254. Hence $\dim \mathrm{Ker}\ d_1 = \dim \mathrm{Coker}\ \delta_1 = h^n(\Sigma_Y)$. The proof of theorem 2 is now complete.

□

REMARK 1. The case of cohomology in the "middle dimension" n is the most important one, because Lefschetz' theorem compares other cohomology groups with middle-dimensional cohomology of linear sections.

REMARK 2. In the case of relative curves $(n=1)$, one has the formulae $h^n(\Sigma_Y) = \#\{\text{double points}\} - \#\{Y_i\} + 1 = \mathrm{genus}(X_\eta) - \Sigma\ \mathrm{genus}\ Y_i$. Moreover the locally invariant periods are all in M_0 iff there is strong degeneration at s_0 .

§ 5. CONCLUSION

Having now theorems 3.2 and 4.2 at hand, we are ready to give a partial answer to Grothendieck's conjecture 2.2 concerning periods of varieties which lie "near" a strong degeneration (3.2). This deduction is made possible using theorem of chapter VII on diophantine properties of values of G-functions. We shall not reach however a full algebraic independence result, but algebraic independence in "bounded degree δ " : relations of degree $\leq \delta$ between "nice" periods come from Hodge cycles if the variety is sufficiently close (depending on δ) to the degeneration in the family. Here is a precise statement.

Let S' denote as before a smooth connected curve (not necessarily complete) over a number field $K_0 \subset \mathbb{C}$, and let S be the complement of a point $s_0 \in S'(K_0)$. We write $d(.,.)$, resp. $h(.)$, for the distance defined by the "sup norm", resp. the height, with respect to some projective embedding of S' ; e.g. for S' = open subset of $\mathbb{A}^1_{K_0}$, d = Euclidean distance and h = usual height on K_0 . Remember from 4.2 the notion of M_0-period.

MAIN THEOREM. Assume that $X \to S$ is either an Abelian scheme with (completely) multiplicative reduction at s_0 , or resp. a projective smooth morphism of relative dimension n having strong degeneration at s_0 . To these data, one can attach two (theoretically effective) constants $\kappa_1 > 0$ and κ_2, $0 < \kappa_2 < 1$, with the following property. Let $\delta > 1$, and let $s \in S(K)$ (for any number field $K \subset \mathbb{C}$) such that

$$h(s) \geq \delta^{\kappa_1} \text{ and } d(s,s_0) \leq \exp(-[K:\mathbb{Q}]\delta^{\kappa_1}h(s)^{\kappa_2}) ;$$

then every polynomial relation of degree $\leq \delta$ with coefficients in K among the values at s of the locally invariant relative 1-periods (resp. M_0-n-periods) comes from Hodge cycles on X_s ; more precisely, such a relation comes from specializations at s of relative Hodge cycles.

Proof: We may assume that $K = K_0$ after extending K_0, K if necessary. Let us first choose a local parameter x on S', centered at s_0. Changing the local bases which enter the definition of relative periods does not affect the result. Hence in accordance with theorem 4.2.1 resp. 4.2.2 (with $\nu=2$ or $n+1$), we may assume that the x-expansions of the relevant locally invariant relative (resp. M_0-n)-periods are G-functions, say $y_1, y_2, \dots, y_\mu$ (where μ = half square of the 1st Betti number (= 2n) in the Abelian scheme case, or $h^n(\Sigma_Y)$ times the nth Betti number resp. in the strongly degenerating case). We put $y_0 = 1$, and consider some homogeneous polynomial of degree δ, say $q \in K[x_0, \dots, x_\mu]$, satisfying $q(y_0, \dots, y_\mu)(\xi) = 0$ for $\xi = x(s)$. Note further that by virtue of the general theory of heights [45], the conditions made upon $h(s)$ and $d(s, s_0)$ correspond to similar conditions upon $h(\xi)$ and $|\xi|$. Now for suitable constants κ_1 and κ_2, the conclusion of theorem VII 4 applies, showing thereby that q defines no strongly non-trivial relation. This means accordingly that the hypersurface $\Sigma(q)$ of $\mathbb{P}^\mu_K = \operatorname{Proj} K[x_0, \dots, x_\mu]$ defined by q contains some irreducible component of the specialization at $x = \xi$ of the following closed subscheme of $\mathbb{P}^\mu_{K[x]_{(x-\xi)}}$:

$$\widetilde{\Theta} := \operatorname{Proj}\Big(K[x]_{(x-\xi)}[x_0, \dots, x_\mu] \Big/ \text{homogeneous } \operatorname{Ker}\Big(K[x]_{(x-\xi)}[x_0, \dots, x_\mu] \longrightarrow K[x]_{(x-\xi)}[y_0, \dots, y_\mu]\Big)\Big).$$

According to corollary 3.4, $\widetilde{\Theta} \otimes_{K[x]_{(x-\xi)}} \bar{\mathbb{Q}}(S) \supset \prod_{X/S}$ (notation of remark 3.4.2). In particular the affine part of $\Sigma(q)_{\bar{\mathbb{Q}}}$ in $\mathbb{A}^\mu_{\bar{\mathbb{Q}}}$ contains an irreducible complement of $\prod_{X_s}$, that is, the period relation $q\left(\frac{1}{(2i\pi)^{(1 \text{ or } n)}}\int_{\gamma_1(s)} \omega_1(s), \dots, \frac{1}{(2i\pi)^{(1 \text{ or } n)}}\int_{\gamma_{b^1/2 \text{ or } h^n(\Sigma_Y)}(s)} \omega_{b^1 \text{ or } b^n}(s)\right) = 0$ comes from Hodge cycles.

□

REMARK 1. With replacing theorem VII 4 by its p-adic variant loc. cit., we obtain a variant of the main theorem in which periods are replaced by the p-adic evaluations at $x = \xi$ of the G-functions which express locally invariant relative periods. It would be very interesting to interpret these p-adic values as p-adic periods on X_s.

REMARK 2. (Case of an Abelian scheme). One may hope weakening the proximity assumption on the point s in the following manner: even if s does not satisfy the hypotheses of the theorem, it may happen that there exists $s' \in S(K')$, such that X_s and $X_{s'}$ are isogeneous (hence have essentially the same periods), but s' satisfies the proximity condition. However, it seems too optimistic to hope reaching transcendence results this way, see ex.2 below.

PROBLEM. We have shown (in chapter V) that all entries of the normalized uniform part of a Picard-Fuchs differential system (in the variable x) associated with $(H^1_{DR}(X/S),\nabla)$ are G-functions. In the situation of theorem 4.2.1 (see 4.2.2), how can these G-functions be related to the (not necessarily locally invariant) periods? A precise answer to this question would give rise to a variant of our main theorem in which all periods of X_s occur: under the proximity hypothesis on s, every ideal of relations (with coefficients in $K(2i\pi)$ among the periods of X_s) of height $\geq$ (?) should contain a relation which comes from Hodge cycles.

EXERCISES. 1) Compare the example given in 1.2 to theorems 4.2.1 and 4.2.2 respectively. In particular, express the relative period $y_1 = \frac{\sqrt{-2}}{12}(1-x)^{1/4}\,{}_2F_1(\frac{1}{12}, \frac{5}{12}, 1, x)$ as the diagonal of an algebraic function in two variables.

2) Let $(S'/K_0), s_0)$ be as in the main theorem, and let X/S be an Abelian scheme with multiplicative reduction at s_0. Assume that there exists a sequence $(s_n)_{n>0}$ of points in $S(\bar{\mathbb{Q}}) \subset S(\mathbb{C})$ such that $\forall \varepsilon > 0$, $h(s_n) = 0\left(\frac{\log d(s_n,s_0)^{-1}}{[K_0(s_n):K_0]}\right)^{1+\varepsilon}$, and the fibers X_{s_n} are isogeneous Abelian varieties of CM type. Show that each fiber X_{s_n}, $n > 0$, satisfies the Grothendieck conjecture. Then test (?) the assumption on the elliptic pencil presented in (1.2.3). (Hint: use the Fourier expansion of J in order to obtain $\log d(s_n,s_0)^{-1} \sim \log c_n$, the conductor of the order $\mathcal{O}_n$ of complex multiplication of X_{s_n}; on the other side $[K_0(s_n):K_0] \approx$ class

number of $\mathcal{O}_n$, and see [44] for an estimate in terms of c_n , due to G. Shimura - in particular $[K_0(s_n):K] \ll \varphi(c_n)$; at last, study $h(s_n)$ by comparing it to Faltings' modular height.)

3) (For a zealous reader!) Extend results of this chapter to the case of Poincaré normal functions (in particular, "logarithms" of sections of an Abelian scheme).

APPENDIX: SPECIAL MUMFORD-TATE GROUPS AND ABSOLUTE HODGE CYCLES

Let X be a complete smooth complex algebraic variety, and let $t \in H^n_{DR}(X)^{\otimes m}$ for some $m,n \in \mathbb{N}$.
To any automorphism $\sigma : \mathbb{C} \longrightarrow \mathbb{C}$, there is an associated "conjugate cycle" t^σ lying in $H^n_{DR}(X^\sigma)^{\otimes m}$.

DEFINITION. The cycle t is an <u>absolute Hodge cycle</u> iff for every automorphism $\sigma : \mathbb{C} \longrightarrow \mathbb{C}$, t^σ is a Hodge cycle, that is $t^\sigma \in Hg^n(X^\sigma)$.
It is believed that every Hodge cycle should be absolute; in other words, for every conjugation σ, $t \longmapsto t^\sigma$ should induce a bijection $Hg^n(X) \simeq Hg^n(X^\sigma)$; this is "Deligne's hope", which is a weakened form of the Hodge conjecture asserting the algebraicity of Hodge cycles.
The main result of [23]I (which we shall not use) is a proof of this conjecture for Abelian varieties; in fact [23]I offers a stronger result involving l-adic cohomologies as well. For us, the most relevant property of absolute Hodge cycles is the following one:

PROPOSITION (Deligne, loc. cit.). <u>If</u> X <u>is defined over an algebraically closed subfield</u> k <u>of</u> $\mathbb{C}$, <u>every absolute Hodge cycle</u> $t \in Hg^n(X_{\mathbb{C}})$ <u>descends to a cycle</u> $t_k \in Hg^n(X_k)$. <u>In particular if every Hodge cycle on</u> $X_{\mathbb{C}}$ <u>is absolute, the natural injection</u> $Hg^n(X_k) \hookrightarrow Hg^n(X_{\mathbb{C}})$ <u>is a bijection</u>.

This is a nice situation because Hodge groups of complex varieties admit a convenient description via Mumford-Tate groups, as follows (see also [23]I). Recall that a Hodge structure on a $\mathbb{Q}$-vector space V is determined by its weight and a morphism of real affine algebraic groups:

$S^1 \xrightarrow{h} GL(V \otimes_{\mathbb{Q}} \mathbb{R})$, where $S^1 = \{z \in \mathbb{C}, |z| = 1\}$; $h(z)$ acts as $z^p \bar{z}^q$ on the (p,q)-component.

DEFINITION. The <u>special Mumford-Tate group</u> of the Hodge structure on V is the smallest algebraic subgroup of $GL(V)$ defined over $\mathbb{Q}$ and containing $h(S^1)$. We denote by $G^n_{smt}(X)$ the special Mumford-Tate group of $V = H^n(X^{an},\mathbb{Q})$.

PROPOSITION. <u>The isomorphism</u> $P^n_{X/\mathbb{C}} : H^n_{DR}(X) \xrightarrow{\sim} H^n(X^{an},\mathbb{Q}) \otimes_{\mathbb{Q}} \mathbb{C}$ <u>induces an isomorphism of algebraic groups</u> $G^n_{shg}(X) \simeq G^n_{smt}(X)|_{\mathbb{C}}$, see D. Mumford, "Families of Abelian varieties", Algebraic Groups and Discontinuous Subgroups, Proc. Symp. Pure Math. vol.9 A.M.S. Providence R.I. 1966, 347-351, or [23]I. Using the polarizability of the Hodge structure $H^n(X^{an},\mathbb{Q})$, one can show that $G^n_{smt}(X)$ is <u>a connected reductive group</u> (in particular, if we assume Deligne's hope this shows that it was unnecessary to complicate the definition of "relations coming from Hodge cycles" as we did in 2.2 by considering connected components ...).
Let us describe this group more closely in case X is a complex <u>Abelian variety</u>, see e.g. [23 1/2] or [50]4.
The group $G = G^1_{smt}(X) \subset GL(V)$ satisfies the following properties:

i) $End_G V \simeq End_0 X := (End\ X) \otimes_{\mathbb{Z}} \mathbb{Q}$,

ii) for each polarization of X , G preserves the corresponding skew-symmetric form $\langle\ ,\ \rangle$ on V .

iii) assume that X is simple of dimension g , such that $\dim_{\mathbb{Q}} End_0 X = ef^2$ where $e = [Z:\mathbb{Q}]$ and Z is the center of $End_0 X$. Then $V_{\mathbb{C}} := V \otimes_{\mathbb{Q}} \mathbb{C}$ splits into a direct sum of $G_{\mathbb{C}}$-primary summands: $V_{\mathbb{C}} = \bigoplus_{\sigma: Z \hookrightarrow \mathbb{C}} V_\sigma$ and $Hom_{G_{\mathbb{C}}}(V_\sigma, V_\tau) = 0$ for $\sigma \neq \tau$; the groups $pr_{V_\sigma} G_{\mathbb{C}}$ are isomorphic. The decompositions of the primary components into sums of $G_{\mathbb{C}}$-irreducible components $V_\sigma = \underbrace{W_\sigma \oplus \ldots \oplus W_\sigma}_{f \text{ factors}}$ are as follows, according to the four cases in Albert's classification of endomorphism algebras (D. Mumford "Abelian varieties" Tate Inst. Fund. Res. Bombay, and Oxford Univ. Press, London 1970) § 21:

$$\left.\begin{array}{r}\text{Type I}: f = 1,\ \varepsilon = -1\\ \text{II}: f = 2,\ \varepsilon = -1\\ \text{III}: f = 2,\ \varepsilon = +1\end{array}\right\}\ G_{\mathbb{C}} \text{ is semi-simple,}$$

IV : $W_{\bar\sigma}$ is the representation of $G_{\mathbb{C}}$ dual to W_σ , and the center of $pr_{W_\sigma} G_{\mathbb{C}}$ is contained in the homothety group. Here the symbol ε means that on W_σ there exists a $G_{\mathbb{C}}$-invariant symmetric $(\varepsilon = +1)$ or skew-symmetric $(\varepsilon = -1)$ bilinear form.

iv) the semi-simple part, say $\mathfrak{g}$, of the Lie algebra of $G_{\mathbb{C}}$ is a classical semi-simple Lie algebra, and its faithful representation $V_{\mathbb{C}}$ is generated by microweights.
This means that the simple factors ρ_i of any of the absolutely irreducible representation $\rho = \rho_1 \otimes \ldots \otimes \rho_n$: $\mathfrak{g} = \mathfrak{g}_1 \times \ldots \times \mathfrak{g}_n \longrightarrow \text{End } W_\sigma$, with $W_\sigma = W_1 \otimes \ldots \otimes W_n$, belong to the following table:

Type	Microweights ω	Degree of ρ_ω	$\varepsilon(\omega)$ (if any)
A_l $l \geq 2$	$\omega_1, \ldots, \omega_{\frac{l+1}{2}}, \ldots, \omega_l$	$\binom{l+1}{1}, \ldots, \binom{l+1}{\frac{l+1}{2}}, \ldots, \binom{l+1}{l}$	$\varepsilon\left(\omega_{\frac{l+1}{2}}\right) = (-1)^{\frac{l+1}{2}}$ when l is odd
B_l $l \geq 2$	ω_l	2^l	$\varepsilon = (-1)^{l(l+1)/2}$
C_l $l \geq 2$	ω_1	$2l$	$\varepsilon = -1$
D_l $l \geq 3$	$\omega_1, \omega_{l-1}, \omega_l$	$2l, 2^{l-1}, 2^{l-1}$	$\varepsilon(\omega_1) = 1$ $\varepsilon(\omega_{l-1}) = \varepsilon(\omega_l) = (-1)^{l/2}$ when l is even

In this table, the symbol ρ_ω denotes the irreducible representation with highest weight ω ; the symbol $\varepsilon(\omega)$ means that for ρ_ω there exists an invariant bilinear form $(\varepsilon = \pm 1)$ (remember that if W_σ carries an invariant bilinear form $\langle\, ,\rangle_\sigma$, the form is either symmetric $(\varepsilon = 1)$ or skew-symmetric $(\varepsilon = -1)$, and each component W_i carries an invariant bilinear form $\langle\, ,\rangle_i (\varepsilon_i = \pm 1)$, with $\langle\, ,\rangle_\sigma = \langle\, ,\rangle_1 \otimes \ldots \otimes \langle\, ,\rangle_n$ and $\varepsilon = \varepsilon_1 \ldots \varepsilon_n$).

Granting properties i)...iv), it is often easy to determine $G^1_{smt}(X)$, see e.g. next chapter.
At last it is perhaps worth noting that the notions of Hodge cycles and groups still exist for non-proper varieties, or for 1-motives, see [3]. This allows to generalize the Grothendieck conjecture. In particular, we leave to the reader to check the following fact: the Grothendieck conjecture about the periods of the 1-motive over $\bar{\mathbb{Q}}$ $[\alpha^{\mathbb{Z}} \longrightarrow \mathbb{G}_m]$ amounts to the transcendence of $\frac{\log \alpha}{2i\pi}$ for α non-torsion in $\bar{\mathbb{Q}}^{\times}$ (Gel'fond).

Chapter X Endomorphisms in the Fibers of an Abelian Pencil

§ 1. INTRODUCTION: DISTRIBUTION OF THE EXCEPTIONAL FIBERS

1.1 Let $X \to S$ be a one-parameter algebraic family of Abelian varieties. We denote by D the endomorphism algebra $(\mathrm{End}\, X_{\bar{\eta}}) \otimes_{\mathbb{Z}} \mathbb{Q}$ of the geometric generic fibers
Then for any geometric point s of S, there is a natural embedding of $\mathbb{Q}$-algebra (with unit) $D \hookrightarrow \mathrm{End}_0 X_s := (\mathrm{End}\, X_s) \otimes_{\mathbb{Z}} \mathbb{Q}$. The present chapter is devoted to a study of the set of fibres X_s such that this inclusion is <u>not</u> an equality: $D \neq \mathrm{End}_0 X_s$. We shall refer to such fibres as the <u>exceptional</u> fibers.

1.2 The nature of the problem, and what is to be expected, is already well illustrated by the case of an elliptic pencil, say the one considered in IX (1.2.3). Here the exceptional fibers are the so-called fibers of C.M. type, i.e. elliptic curves which admit the complex multiplication by an order R in some imaginary quadratic field $\mathbb{Q}(\sqrt{-D})$, where $-D$ = discriminant of R. There are infinitely many such fibers X_s, when s runs over $S(\bar{\mathbb{Q}})$. However, if we bound the degree of the residual field of s, say $[\kappa(s):\mathbb{Q}] \leq H$, it then happens that there remain only finitely many exceptional fibers. Indeed, it is well-known that $[\kappa(s):\mathbb{Q}] = [\mathbb{Q}(J):\mathbb{Q}]$ is just the class number of R. Thus our claim of finiteness corresponds to the classical fact that there are only finitely many orders of complex multiplication with class number $\leq H$, cf [44]. We shall see that this special example reflects a generality. Our method is a continuation of the method of the last chapter; the new tool is the construction of non-trivial global relations between the values at any exceptional fiber of the locally invariant periods around a degeneration (under suitable hypotheses). Here is the precise statement.

1.3 Let S' a smooth connected (not necessarily complete) curve over a number field K, and let S be the complement of a closed point s_0 in S'. Let h denote a Weil height on S'.

THEOREM. *Let* $f : X \longrightarrow S$ *be an Abelian scheme with multiplicative reduction at* s_0 *(i.e. the fiber at* s_0 *of the connected Néron model is a torus). Assume that the geometric generic fiber* $X_{\bar\eta}$ *is a simple Abelian variety of odd dimension* $g > 1$. *Then there are only a finite number of fibers* X_s *with bounded residual degree (say* $[K(s):K] \leq d$ *), such that there is no ring embedding* $\mathrm{End}\, X_s \hookrightarrow M_g(\mathbb{Q})$. *More precisely, the height* $h(s)$ *is bounded effectively by a power of* $d+1$.

REMARK 1. The first conclusion follows from the second one via Northcott's theorem [45].

REMARK 2. Because the fiber at s_0 of the connected Néron model is a torus, say X_{s_0}, we have $\mathrm{End}_0 X_{\bar\eta} \subset \mathrm{End}_0(X_{s_0}|\bar{\mathbb{Q}}) \simeq M_g(\mathbb{Q})$ by functoriality. Hence any fiber such that $\mathrm{End}\, X_s \not\hookrightarrow M_g(\mathbb{Q})$ is exceptional. On the other hand, an easy specialization argument taking into account the fact that $\dim S = 1$, shows that if X_s is an exceptional fiber for some $s \in S(\mathbb{C})$, then in fact $s \in S(\bar{\mathbb{Q}})$.

REMARK 3. It is a pity that the case of an elliptic pencil is ruled out in the theorem! Indeed, the conclusion of the theorem would provide in this case (in accordance with the above discussion 1.2.) an effective version of Siegel's theorem $\log\sqrt{D} = O(\log H(-D))$ (in fact Siegel's theorem states $\log\sqrt{D} \sim \log H(-D)$). The point is that by looking at the Fourier expansion of the modular form $J(\tau)$, one can easily write down a lower bound for $h(J)$ of the shape: $h(J) \geq \frac{\sqrt{D}}{H(-D)} - O(1)$, at least when R is the maximal order; from this starting point, one may reverse the argumentation of 1.2.

1.4 At the cost of replacing K by a finite extension, and S by an étale neighbourhood, we may assume that $\mathrm{End}_0 X_{\bar\eta} = (\mathrm{End}_S X) \otimes_{\mathbb{Z}} \mathbb{Q}$, (which we shall denote by D), that $H^1_{DR}(X/S)$ is free, and that f is projective. As we have remarked earlier, $D \hookrightarrow M_g(\mathbb{Q})$.

Also we may assume that the set of fibers X_s being considered in the theorem is non-empty. We put $\hat{D} = \mathrm{End}_0 X_s$; its center $\hat{Z}$ is a product of fields $\hat{Z}_i$ and $\hat{D}$ splits accordingly into a product of simple central $\hat{Z}_i$-algebras $\hat{D}_i$. For each $\hat{D}_i$, we choose a maximal subfield $\hat{E}_i$ (we shall precise this choice later). We put:

$$E = \prod \hat{E}_i$$

$$\hat{F}_i = \text{Galois closure of } \hat{E}_i \text{ in } \mathbb{C}$$

$$\hat{F} = \text{Compositum of the } \hat{F}_i \text{ in } \mathbb{C}$$

$$\hat{K} = \text{a compositum of } K(s,s_0) \text{ and } \hat{F} .$$

One has $[\hat{F}:\mathbb{Q}] \leq (2g)!$, and $[\hat{K}:\mathbb{Q}] = O(d)$. The theorem follows immediately from the following claim:

(1.4.1) <u>There exists</u> $\kappa > 0$, <u>independent of</u> s <u>and effectively computable, such that</u> $h(s) \leq ([\hat{K}:\mathbb{Q}]+1)^{\kappa}$ <u>for any</u> s <u>as in the theorem</u>.

At the cost of replacing once again the ground field by a finite extension, there exists a rational function, say $1/x \in K(S')$, regular outside s_0 and having a pole at s_0 (Riemann-Roch). We choose its inverse x as a local parameter. Of course this may introduce finite ramification at s_0 , but this does not matter: IX 4.2.1 still applies, and shows the existence of a period matrix (once we have chosen an embedding $i : K \hookrightarrow \mathbb{C}$), such that the Taylor expansion in the variable x of its locally invariant entries around s_0 are G-functions, say $y_1,\dots,y_{2g^2} \in K[[x]]$. We shall show that for appropriate choice of this period matrix, the following lemma holds true:

LEMMA. <u>There exists homogeneous non-trivial global relations of degree</u> $\delta \leq 2[\hat{K}:\mathbb{Q}]$ <u>with coefficients in</u> $\hat{K}$, <u>between the values at</u> $\xi = x(s)$ <u>of</u> $y_1,\dots,y_{2g^2}$; <u>in other words,</u> $\xi \in \underline{|\,|\,|}^{\delta}(y_1,\dots,y_{2g^2})$, <u>with the notation of</u> VII 5.2.

By theorem VII 5.2 (or more accurately theorem E stated in the general introduction), this implies (1.4.1). The next two paragraphs are devoted to a proof of this lemma.

§ 2. PERIOD RELATIONS ON EXCEPTIONAL FIBERS

2.1 Let us start from the general situation of an Abelian scheme $f : X \longrightarrow S$, where S is a smooth connected variety over a subfield k of $\mathbb{C}$, We first introduce some notations:

$D = (\mathrm{End}_S X) \otimes_{\mathbb{Z}} \mathbb{Q}$,

$Z = \prod Z_i$ = center of D

$E = \prod E_i$ = a maximal commutative semi-simple algebra in D

F_i = Galois closure of the field E_i in $\mathbb{C}$

F = compositum of the F_i's .

Replacing if necessary S by an open subset (and K by the finite extension F), let us assume that $H^1_{DR}(X/S)$ is free, and that the $E \otimes_{\mathbb{Q}} k(S)$ - module $\Gamma H^1_{DR}(X/S) \otimes k(S) \simeq H^1_{DR}(X_\eta)$ splits into a direct sum of $k(S)$-spaces with connection $\bigoplus_{\sigma: E \to k(S)} W^\sigma_{DR}$, where the action of E on W^σ_{DR} is through the non-zero morphism of algebras σ (which does not necessarily preserve units). We have: $W^\sigma_{DR} \simeq W^\tau_{DR}$ if $\sigma_{|Z} = \tau_{|Z}$.

Similarly, the splitting $E \otimes_{\mathbb{Q}} F \simeq \bigoplus_{\sigma: E \to \mathbb{C}} F^\sigma$ induce a splitting of local system of F-vector spaces: $R_1 f^{an}_{\mathbb{C}\,*}(F) = \bigoplus_{\sigma: E \to \mathbb{C}} W_\sigma$, where the action of E on W_σ is through σ . We have: $W_\sigma \simeq W_\tau$ if $\sigma_{|Z} = \tau_{|Z}$.

Moreover each $W_\sigma \otimes_F \mathbb{C}$ is naturally bigraded, because E acts as morphisms of variation of $\mathbb{Q}$-Hodge structure. Via the isomorphism $P^1_{X/S}$, $W_{DR} \otimes_{k(S)} M_V$ is dual to $W_\sigma \otimes_F M_V$; here M_V denotes the field of meromorphic functions on some open subset V of $S^{an}_{\mathbb{C}}$, such that $R_1 f^{an}_{\mathbb{C}\,*}(F)_{|V}$ admits a trivialization (= frame). Choosing a frame $\{\gamma_{\tau,j}\}$ adapted to the splitting $(\oplus W_\tau)_{|V}$, and a basis $\{\omega_{\sigma,i}\}$ for $\Gamma H^1_{DR}(X/S) \otimes k(S)$ inside $\Gamma H^1_{DR}(X/S)$ adapted to the splitting $\bigoplus_\sigma W^\sigma_{DR}$, we obtain the following period relations:

(2.2.1) $$\frac{1}{2i\pi} \int_{\gamma_{\tau,j}} \omega_{\sigma,i} = 0 \quad \text{for } \sigma \neq \tau$$

$$(2.1.2)\qquad \frac{1}{2i\pi}\int_{\gamma_{\sigma,j}} \omega_{\sigma,i} = \frac{1}{2i\pi}\int_{\gamma_{\tau,j}} \omega_{\tau,i} \quad \text{if} \quad \sigma_{|Z} = \tau_{|Z} .$$

REMARK 1. Strictly speaking, these are not relative periods but only F-linear combinations of relative periods. We shall allow however this slight abuse of language.

REMARK 2. Of course all of this holds in particular if S = a point. To distinguish this case, we shall put a roof on the related objects: thus $\hat{D}$, $\hat{Z}$, ..., $\hat{W}_{\hat{\sigma}}$, ..., $\hat{\omega}_{\hat{\sigma},i}$, ...
We shall have to consider the case of a fiber X_s in a family X/S : hence we shall write $D = (\mathrm{End}_S X) \otimes_{\mathbb{Z}} \mathbb{Q}$, $\hat{D} = (\mathrm{End}\, X_s) \otimes_{\mathbb{Z}} \mathbb{Q}$, ... Moreover in this situation, we choose E, $\hat{E}$ such that $E \subset \hat{E}$; this implies some splittings of the type $(W_\sigma)_s \otimes_F \hat{F} = \bigoplus_{\substack{\hat{\sigma}:\hat{E} \to k(S) \\ \hat{\sigma}_{|E} = \sigma}} \hat{W}_{\hat{\sigma}}$, etc. ...

2.2 LEMMA. <u>If the generic fiber X_η is absolutely simple of odd dimension g , if moreover S is a curve and X/S has multiplicative reduction at some point of $\bar{S} \setminus S$, then $D = Z = E$ is a totally real field</u>.

Proof: the first assumption leaves only type I or IV in Albert's classification of endomorphism algebras (see ch. IX, app.). We have to show, under both assumptions, that $X_{\bar{\eta}}$ cannot be of type IV. As a Wronskian of a ∇-module arising from geometry, Det W^σ_{DR} satisfies: for some positive integer n, $(\mathrm{Det}\, W^\sigma_{DR})^{\otimes n}$ is trivial as a ∇-module, see e.g. VIII ex.6. Via $P^1_{X/S}$, we get: the local system $(\mathrm{Det}(W_\sigma \otimes_F \mathbb{C}))^{\otimes n}$ is constant. Moreover, the second assumption implies, via Mustafin's theorem (IX 3), that $(\mathrm{Det}(W_\sigma \otimes_F \mathbb{C}))^{\otimes n}$ has type (p,p) ; whence:

(2.2.1) the $(0,-1)$ and $(-1,0)$ - components of $W_\sigma \otimes_F \mathbb{C}$ have the same dimension. In particular $\dim_F V_\sigma$ is even. On the other side, the number of pieces V_σ is divisible by $[Z:\mathbb{Q}]$, which is even for type IV. Thus type IV is ruled out by our assumption: g is odd.

□

It follows from this lemma that any polarization, which corresponds to a certain skew-symmetric form $2i\pi\langle\ ,\ \rangle$ on $R_1 f^{an}_{\mathbb{C}*}\mathbb{Q}$ with values in $\mathbb{Q}(1)$, induces a skew-symmetric form $2i\pi\langle\ ,\ \rangle_\sigma$ on W_σ with values in $F(1)$ (a morphism of local system), see IX app.

2.3 In the sequel, S is again the complement of a closed point s_0 in a k-curve S', and the X/S is an Abelian scheme of odd relative dimension g, with multiplicative reduction at s_0. Let us consider the inclusion of a unit disk Δ in $S'^{an}_{\mathbb{C}}$, such that 0 maps to s_0, assuming for simplicity that s_0 is k-rational (as for the open subset V of $S^{an}_{\mathbb{C}}$ on which the periods are defined, we may choose an open dense subset of the image of Δ^*). We have seen in the previous chapter that the maximal constant sub-system of $(R_1 f_\Delta)_*(\mathbb{Q})$ has dimension g. After tensoring by F, this sub-system splits into $\bigoplus_{\sigma:E\to\mathbb{C}} W^1_\sigma$, where W^1_σ is a constant sub-system of $W_{\sigma|\Delta}$ of half dimension.

LEMMA. W^1_σ <u>is maximal totally isotropic with respect to</u> $<\ ,\ >_\sigma$.

Proof: for any $z \in \Delta^*$, let us consider the logarithm $2i\pi N^*_z$ of the local monodromy on $((R_1 f_\Delta)_*(\mathbb{Q}))_z$. It induces a F-linear map $2i\pi N^\sigma_z$ on $(W_\sigma)_z$. Under the multiplicative reduction assumption, $(W^1_\sigma)_z$, which is $\operatorname{Ker} N^\sigma_z$, has half dimension, say n. Because $(N^\sigma_z)^2 = 0$, we also have $\operatorname{Im} 2i\pi N^\sigma_z = (W^1_\sigma)_z$ for dimension reasons. Because $<\ ,\ >_\sigma$ is a morphism of local system $W_\sigma \otimes W_\sigma \longrightarrow F$, one has

$$<2i\pi N^\sigma_z(V),w>_\sigma + <v,2i\pi N^\sigma_z(w)>_\sigma = 0 \quad \text{for every } v,w \in (W_\sigma)_z .$$

This shows that $(W^1_\sigma)^\perp_z \subset \operatorname{Im} N^\sigma_z = (W^1_\sigma)_z$. Since $\dim(W^1_\sigma)_z = \frac{1}{2}\dim(W_\sigma)_z$, we are done.

□

On the other side, the polarization (Poincaré-) dual to the one considered before corresponds to a skew-symmetric form $<\ ,\ >_{DR}$ on $\Gamma H^1_{DR}(X/S) \otimes k(S)$ with values in $k(S)$, which induces a skew-symmetric form $<\ ,\ >^\sigma_{DR}$ on W^σ_{DR} by virtue of lemma 2.2. With arranging our bases $\omega_{\sigma,i}$, resp. $\gamma_{\sigma,j}$, we may represent $<\ ,\ >^\sigma_{DR}$ as well as $<\ ,\ >_\sigma$ by the matrix $J = \begin{pmatrix} 0 & -I \\ I & 0 \end{pmatrix}$, and assume that $\gamma_{\sigma,1},\dots,\gamma_{\sigma,n} \in W^1_\sigma$, i.e. are locally invariant, for $n = g/_{[E:\mathbb{Q}]} = \frac{1}{2}\dim W_\sigma$.

The relative period matrix is constituted by "diagonal" blocks of the same size n, indexed by $\sigma : D = E \longrightarrow \mathcal{M}_V$ (which contains both $k(S)$ and $\mathbb{C}$):

(2.3.0)

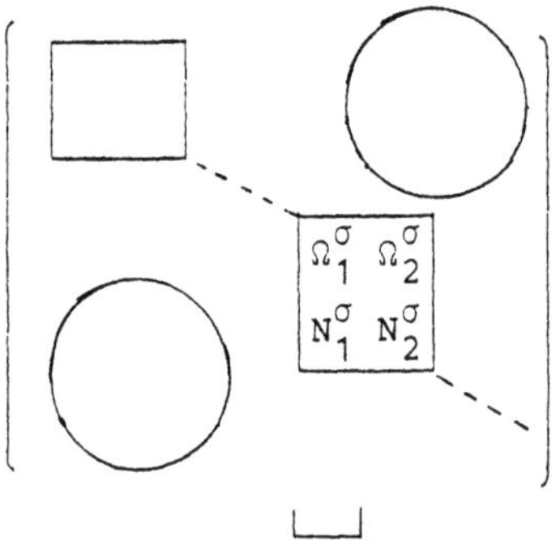

locally invariant part.

In fact, it is classical that the subspace $\Gamma(e^*\Omega^1_{X/S}) \otimes k(S) \simeq \simeq F^1H^1_{DR}(X_\eta)$ is maximal totally isotropic with respect to $<\ ,\ >_{DR}$ (e = neutral section). Moreover, it follows from (2.2.1) that $F^1W^\sigma_{DR} = \Gamma(e^*\Omega^1_{X/S}) \otimes k(S) \cap W^\sigma_{DR}$ is maximal totally isotropic in W^σ_{DR} with respect to $<\ ,.>^\sigma_{DR}$. Therefore we may select $\omega_{\sigma,1},\dots,\omega_{\sigma,n}$ in $F^1W^\sigma_{DR}$, i.e. to be forms of the "first kind", whence the above use of the familiar notation Ω_1,Ω_2.

REMARK. In the classical language, the label "periods" is reserved for the entries of the matrices $2i\pi\Omega^\sigma_{1\text{ or }2}$, while the entries of $2i\pi N^\sigma_{1\text{ or }2}$ are called "pseudo-periods".
The duality between $<\ ,\ >^\sigma_{DR}$ and $2i\pi<\ ,\ >_\sigma$ (expressed via $P^1_{X/S}$), once being translated in terms of periods, gives the famous "Riemann relations":

$$\begin{pmatrix} {}^tN_1^\sigma \\ {}^tN_2^\sigma \end{pmatrix} (\Omega_1^\sigma\ \Omega_2^\sigma) - \begin{pmatrix} {}^t\Omega_1^\sigma \\ {}^t\Omega_2^\sigma \end{pmatrix} (N_1^\sigma\ N_2^\sigma) = \frac{J}{2i\pi},$$

that is to say:

(2.3.1) $\quad {}^tN_1^\sigma\,\Omega_1^\sigma$ is symmetric,

(2.3.2) $\quad {}^tN_2^\sigma\,\Omega_2^\sigma$ is symmetric,

(2.3.3) $\quad {}^tN_2^\sigma\,\Omega_1^\sigma - {}^t\Omega_2^\sigma\,N_1^\sigma = \frac{I}{2i\pi}$.

Let us illustrate this by the elliptic example IX(1.2.3): in this case, $E = \mathbb{Q}$ and we may take $\Omega_{1,2} = y_{1,2}$,

$N_{1,2} = 72\,x\,\frac{dy_{1,2}}{dx}$, and (2.3.3) is just the Wronskian relation $y_1 \frac{dy_2}{dx} - y_2 \frac{dy_1}{dx} = \frac{1}{144\, i\pi x}$.

2.4 We keep on our conventions at the beginning of 2.3. Let $s \in S(k)$ be such that the corresponding point on $S^{an}_{\mathbb{C}}$ belongs to the open subset V . In accordance with 2.1 (remark 2), we shall affect symbols associated with X_s with a roof, in order to distinguish them from the relative analogs. We put $\hat{k} = k\hat{F}$, so that there is a splitting $H_{DR}(X_s) \otimes_k \hat{k} = \bigoplus_{\substack{\hat{\sigma}:\hat{E}\to\mathbb{C} \\ \hat{\sigma}\neq 0}} \hat{W}^{\hat{\sigma}}_{DR}$.

CONSTRUCTION(2.4.1). Assume that X_s is an exceptional fiber. We shall construct relations of degree ≤ 2 with coefficients in $\hat{k}(2i\pi)$ among the values at s of the (locally invariant) entries of Ω_1^σ, N_1^σ , for any $\sigma : E \longrightarrow \mathbb{C}$. Moreover, if $g > 1$, it is possible to eliminate $2i\pi$ between these relations and obtain (homogeneous) linear or quadratic period relations with coefficients in $\hat{k}$ (involving possibly several σ's).

If $E = \hat{E}$, then $\hat{D}$ is a simple algebra, $\neq E$ since X_s is exceptional; in particular X_s is not of type I; because E is totally real, X_s is not of type IV, and because g is odd, types II and III are ruled out too. Therefore $E \neq \hat{E}$. We then distinguish three cases.

Case 1: in the splitting $(W_\sigma)_s \otimes_F \hat{F} = \bigoplus_{\hat{\sigma}|\sigma} \hat{W}_{\hat{\sigma}}$, there exists some $\hat{\tau}|\sigma$ such that $\dim_{\hat{F}} \hat{W}_{\hat{\tau}} < \frac{1}{2} \dim_F (W_\sigma)_s$. Then because $\dim_F (W^1_\sigma)_s = \frac{1}{2} \dim (W_\sigma)_s$, we have:

$$\Big(\bigoplus_{\substack{\hat{\sigma}|\sigma \\ \hat{\sigma}\neq\hat{\tau}}} \hat{W}_{\hat{\sigma}}\Big) \cap [(W^1_\sigma)_s \otimes_F \hat{F}] \neq 0 .$$

Let $\gamma \in \Gamma(V, (W_\sigma \otimes_F \hat{F}))$ be a non-zero cycle whose fiber at s belongs to that non-zero space. After 2.1, we have $\frac{1}{2i\pi} \int_{\gamma(s)} \hat{\omega} = 0$, for any $\hat{\omega} \in \hat{W}^{\hat{\tau}}_{DR}$. With expressing $\hat{\omega}$, resp. $\gamma(s)$, as $\hat{k}$-linear combination of the values at s of the $\omega_{\sigma,i}$, resp. $\gamma_{\sigma,j}$ $j = 1,\ldots,n$ (the locally invariant part of the basis), we find a linear relation as wanted.

Case 2: $(W_\sigma)_s \otimes_F \hat{F} = \hat{W}_{\hat{\sigma}_1} \oplus \hat{W}_{\hat{\sigma}_2}$, each $\hat{W}_{\hat{\sigma}_i}$ $(i = 1,2)$ having dimension n , and $\hat{W}_{\hat{\sigma}_2} \cap [(W^1_\sigma)_s \otimes_F \hat{F}] \neq 0$.

Let $\gamma \in \Gamma(V,(W^1_\sigma)_s \otimes_F \hat{F})$ be a non-zero cycle whose fiber at s belongs to that non-zero space. After 2.1, $\frac{1}{2i\pi}\int_{\gamma(s)} \hat{\omega} = 0$ for any $\hat{\omega} \in \hat{W}^{\hat{\sigma}_1}_{DR}$, and this gives a linear period relation as before.

Case 3: $(W_\sigma)_s \otimes_F \hat{F} = \hat{W}_{\hat{\sigma}_1} \oplus \hat{W}_{\hat{\sigma}_2}$, $\dim \hat{W}_{\sigma_i} = n$ for $i = 1,2$, but $\hat{W}_{\hat{\sigma}_2} \cap [(W^1_\sigma)_s \otimes_F \hat{F}] = 0$. (Note that if this situation occurs for one σ , it occurs for all of them by conjugation).
At the cost of replacing F by $\hat{F}$, we may choose the (non-locally invariant) cycles $\gamma_{\sigma,n+1} \cdots \gamma_{\sigma,2n}$ in our basis, such that their fibers at s span $\hat{W}_{\hat{\sigma}_2}$. By substituting to $\{\omega_{\sigma,i}\}$ a basis of $W^\sigma_{DR} \otimes_k \hat{k}$ inside $\Gamma H^1_{DR}(X/S) \otimes_k \hat{k}$, whose fibers at s are given by $\hat{\omega}_{\hat{\sigma}_{1,i}}$ for $i = 1,\dots,n$ and $\hat{\omega}_{\hat{\sigma}_{2,i-n}}$ for $i = n+1,\dots,2n$, we obtain the period relations:

$$(2.4.2) \qquad \begin{cases} \Omega^\sigma_2(s) = 0 \\ N^\sigma_1(s) = N^\sigma_2(s)\,{}^tT \quad \text{for some } T \in M_n(\hat{k}) . \end{cases}$$

Combining this with the Riemann relations (2.3.3), we get:

$$(2.4.3) \qquad {}^tN^\sigma_1(s)\Omega^\sigma_1(s) = \frac{T}{2i\pi} .$$

In terms of the original periods, we get however a slightly more complicated relation, of the shape

$$(2.4.4)_\sigma \qquad ({}^t\Omega^\sigma_1 A + {}^tN^\sigma_1 B)(A'\Omega^\sigma_1 + B'N^\sigma_1)(s) = \frac{T}{2i\pi} , \text{ with } A,B,A',B',T \in M_n(\hat{k}) .$$

Now if $g > 1$, then $n > 1$ or there exists at least two distinct σ's ; thus it is possible to eliminate $2i\pi$ between relations $(2.4.4)_\sigma$, as wanted.

□

REMARK. Using the fact that g is odd, it is possible to show that in case 3, $\hat{E}$ is a field, namely a totally imaginary quadratic extension of E .

§ 3. CONSTRUCTING NON-TRIVIAL GLOBAL RELATIONS

3.1 We go back to the conventions 1.4, and try to make a little more precise our choice of the period matrix (with respect to a complex embedding $\iota : K \hookrightarrow \mathbb{C}$), whose locally invariant part gives rise to the matrix of G-functions

$$Y = \begin{pmatrix} y_1 & \cdots\cdots\cdots & y_g \\ \vdots & & \vdots \\ y_{2g^2-g+1} & \cdots & y_{2g^2} \end{pmatrix} \in M_{2g,g}(K[[x]]) .$$

With no loss of generality, we may assume that s_0 is K-rational. Let $\tilde{S}$ be a regular projective model for S' over $\mathcal{O}_K$, so that we may consider s and s_0 as sections of the arithmetic pencil $\tilde{S} \underset{\mathrm{Spec}\, \mathcal{O}_K}{\times} \mathrm{Spec}\, \mathcal{O}_{\hat{K}} \longrightarrow \mathrm{Spec}\, \mathcal{O}_{\hat{K}}$. From our choice of the local parameter x , it is clear that there exists a system of constants $(\kappa_v)_{v \in \Sigma_f(K)}$, almost all $= 1$, with the following property: $\forall w \in \Sigma_f(\hat{K})$ such that $w|v$, $|x(s)|_w < \kappa_v^{[\hat{K}:\mathbb{Q}]} \Rightarrow s$ and s_0 have the same image in $\tilde{S}(\mathbb{F}_{p(w)})$. We then choose $\zeta \in K^x$ such that $|\zeta|_v \leq \kappa_v$ for all $v \in \Sigma_f(K)$. Multiplying if necessary our local basis of sections of $H^1_{DR}(X/S)$ by $(1-\frac{x}{\zeta})^{-1}$, what amounts to multiplying the y_i's by this factor, we obtain the following extra property:

(3.1.1) let $\xi = x(s)$; if $|\xi|_w < R_w(y_1,\ldots,y_{2g^2})$ for some $w \in \Sigma_f(\hat{K})$, then s and s_0 have the same image in $\tilde{S}(\mathbb{F}_{p(w)})$.

LEMMA. <u>For any other complex embedding</u> $\iota' : K \hookrightarrow \mathbb{C}$, <u>the complex Taylor series</u> $\iota'(y_1),\ldots,\iota'(y_{2g^2})$ (<u>where</u> ι' <u>acts coefficientwise</u>) <u>are again expansions of the locally invariant entries of a period matrix attached to the same basis of local sections of</u> $H^1_{DR}(X/S)$, <u>and to some local frame in</u> $(R_1 f^{an}_{\mathbb{C}\text{ via }\iota'})_*(\mathbb{Q})$.

Proof: indeed, $Y' = \begin{pmatrix} \iota'(y_1) & \cdots\cdots\cdots & \iota'(y_g) \\ & & \\ \iota'(y_{2g^2-g+1}) & \cdots & \iota'(y_{2g^2}) \end{pmatrix}$

satisfies the same Picard-Fuchs differential system as the locally invariant part, say Y_1', of any period matrix as in the statement. Hence, there exists $C \in GL_g(\mathbb{C})$, such that $Y' = Y_1 C$. Now the interpretation of $Y(0)$ as residues on a torus (the s_0-fiber of the neutral component of the Néron model, see IX 4.3) remains valid for $Y'(0)$ by conjugation. Arranging the invariant part of the local frame in $(R_1 f^{an}_{\mathbb{C} \text{ via } \iota'})(\mathbb{Q})$ so that the s_0-fiber (being well-defined, according to IX (4.3.2)) gives the loops involved in residues. With this choice, $C = I$ and we are done.

□

3.2 To any $v \in \Sigma_\infty(\hat{K})$, we associate a complex embedding $\iota_v : \hat{K} \hookrightarrow \mathbb{C}$, and we shall apply the construction (2.4.1) to the exceptional fiber $X_s \underset{\hat{K}}{\times} i_v(\hat{K})$ with $k = \iota_v(K)$.

Let us first remark that we may choose $\omega_{\sigma,i}$, resp. $\gamma_{\tau,j}$ in 2.1, in such a way that they are F-linear combination of our fixed local basis of sections of $H^1_{DR}(X/S)$, resp. of the local frame associated to ι_v and $y_1,\dots,y_{2g^2}$ by the previous lemma.

Construction (2.4.1) furnishes at least a linear or quadratic polynomial q_v, with coefficients in $\hat{K}$ (for $g > 1$), such that $q_v(y_1,\dots,y_{2g})(\xi) = 0$ holds v-adically if $\xi = x(s)$ satisfies

$$|\xi|_v < R_v(y_1,\dots,y_{2g^2}) .$$

We set:

$$(3.2.1) \qquad q = \prod_{\substack{v \text{ such that} \\ |\xi|_v < R_v(y,\dots,y_{2g^2})}} q_v .$$

This is a homogeneous polynomial with coefficients in $\hat{K}$, and degree $\leq 2[\hat{K}:\mathbb{Q}]$ (even $\leq [\hat{K}:\mathbb{Q}]$, if we use the remark at the end of § 2).

REMARK. In general the q_v are distinct polynomials; for relations with coefficients in $\hat{K}(2i\pi)$, this can be observed on CM fibers in elliptic pencils.

3.3 LEMMA. <u>The relation</u> $q(y_1,\dots,y_{2g^2})(\xi) = 0$ <u>is non trivial: it is not the specialization at</u> ξ <u>of any polynomial relation between the series</u> y_i, <u>with coefficients in</u> $\hat{K}[x]_{(x-\xi)}$.

Proof: we fix an embedding $K \hookrightarrow \mathbb{C}$, and consider a non-exceptional fiber X_z, $z \in S(\mathbb{C})$. We refer to IX, app. for what concerns Mumford-Tate groups.

SUBLEMMA (Tankeev). Let X_z be a complex Abelian variety of odd dimension g, such that $E := (\mathrm{End}_0 X_z) \otimes_{\mathbb{Z}} \mathbb{Q}$ is a totally real field. Then up to E-homotheties, there exists a unique polarizing skew-symmetric form $<\,,\,>$ on the E-vector space $V := H^1(X_z^{an}, \mathbb{Q})$, and the special Mumford-Tate group $G_{smt}(X_z)$ may be identified with $\mathrm{Res}_{E/\mathbb{Q}} Sp(V, <\,,\,>)$ (a restriction of scalars à la Weil).

Proof of the sublemma: the assertion about $<\,,\,>$ is classical, and we shall only develop the proof of the second assertion, which amounts to showing that:

i) the projection $\mathfrak{g}$ of the semi-simple complex Lie algebra $\mathfrak{g}_{\mathbb{C}} := \mathrm{Lie}\, G_{smt}(X_z) \otimes_{\mathbb{Q}} \mathbb{C}$ on the direct summand $V_\sigma := V \otimes_{E,\sigma} \mathbb{C}$ of $V \otimes_{\mathbb{Q}} \mathbb{C}$ may be identified with a symplectic Lie algebra $sp(V_\sigma, <\,,\,>_\sigma)$, for every $\sigma : E \longrightarrow \mathbb{C}$.

ii) the embedding $\mathfrak{g}_{\mathbb{C}} \hookrightarrow \prod_\sigma \mathfrak{g}_\sigma$ is an isomorphism.

For i), we look at the table given in the appendix of IX. Because $n = \frac{1}{2} \dim V_\sigma$ is odd, and because V_σ carries an invariant skew-symmetric bilinear form, this forces: $\mathfrak{g}_\sigma$ is simple, of type C_n, and the representation $\rho_\sigma : \mathfrak{g}_\sigma \to \mathrm{End}\, V_\sigma$ is the standard one (here the point is that case A_{4l+1} on the first line of the table is ruled out because $\binom{4l+2}{l+1}^{4l+1}$ is divisible by 4). This proves i).

For ii), we use the classical Goursat lemma: $pr_{\sigma,\tau}\, \mathfrak{g}_{\mathbb{C}}$ is either $\mathfrak{g}_\sigma \times \mathfrak{g}_\tau$ or the graph $\mathfrak{g}_{\sigma,\tau}$ of an isomorphism between $\mathfrak{g}_\sigma$ and $\mathfrak{g}_\tau$. Since the corresponding representation is generated by microweights, and only one such irreducible representation occurs in the table, this isomorphism would necessarily be induced by a $\mathfrak{g}_{\mathbb{C}}$-isomorphism $V_\sigma \simeq V_\tau$, which is ruled out by IX app., point iii). Hence $pr_{\sigma,\tau}\, \mathfrak{g}_{\mathbb{C}} = \mathfrak{g}_\sigma \times \mathfrak{g}_\tau$.

This fact would suffice for our later purpose; however, in order to obtain ii) and by the way the full sublemma, we need only invoke a result of Kolchin-Ribet (see e.g. Ribet "Galois action on division points of Abelian varieties with real multiplications", Am. J. of Math. 98 (1976) 751-804).

□

We now continue the proof of lemma 3.3. According to IX app., there is a canonical isomorphism $G^1_{shg}(X_z) \simeq G^1_{smt}(X_z) \otimes_{\mathbb{Q}} \mathbb{C}$. On the other side, the presence of a fiber with (completely) multiplicative bad reduction in the family implies (via corollary IX 3.4) that the differential Galois group $G^1_{diff}(X_{\hat{K}(S)})$ attached to the Picard-Fuchs differential equation satisfied by "the" relative period matrix contains the generic special Hodge group $G^1_{shg}(X_{\hat{K}(S)})$; in fact $G^1_{diff}(X_{\hat{K}(S)}) = G^1_{shg}(X_{\hat{K}(S)})$ is some inner form of $G^1_{shg}(X_z)$ over $\hat{K}(S)$ - an equality, because G^1_{diff} centralizes E and respects the polarizing form $<,>_{DR}$. It follows that the $\hat{K}(S)$-variety Θ of the coefficients of the relative period matrix (2.3.0) (which is a principal homogeneous space under G^1_{diff}, see IX 3.1), is defined by the ideal generated by the following relations:

$$\left\{\begin{array}{l} \left.\begin{array}{l}(2.3.1)\\(2.3.2)\\(2.3.3)\text{ after elimination of the factor } 2i\pi,\end{array}\right\}\\ \\ \text{vanishing of the non-diagonal blocks in (2.3.0).}\end{array}\right.$$

In particular, Θ is absolutely irreducible and defined over $\mathbb{Q}$. Since the y_i's are K-linear combinations of the coefficients of (2.3.0) expanded into series of x, it follows that the $\hat{K}[x]_{(x-\xi)}$-variety of the y_i's is absolutely irreducible and defined over K; a fortiori. Its specialization at $x = \xi$ is reduced and irreducible. Therefore in order to establish 3.3, it suffices to show that some of the relations q_v is trivial.

If we fix v (and a corresponding embedding $K \hookrightarrow \mathbb{C}$), this amounts to showing that the $\hat{K}$-relations being constructed in (2.4.1) do not belong to the ideal generated by the specialization at s of the functional relations (2.3.1), (2.3.2), and (2.3.3).

Now a simple computation gives that the prime ideal associated with the projection of Θ onto the locally invariant subspace of the space spanned by the Ω_1^σ, N_1^σ, Ω_2^σ, N_2^σ's (for all σ) is generated by the relations (2.3.1).

Therefore lemma 3.3 follows from the fact that the relations constructed in (2.4.1) do not belong to the ideal generated by the specialization at s of the funtional relations (2.3.1). This is easily checked, especially if we substitute once more to Ω_1^σ, N_1^σ suitable $\hat{K}$ - linear combinations and look at the simple relation 2.4.3 (instead of 2.4.4).

3.4 LEMMA. <u>The relation</u> $q(y_1,\ldots,y_{2g^2})(\xi) = 0$ <u>is global if</u> $\mathrm{End}\, X_s \not\hookrightarrow M_g(\mathbb{Q})$.

Proof: it holds v-adically for any $v \in \Sigma_\infty(\hat{K})$ for which $|\xi|_v < R_v(y_1,\ldots,y_{2g^2})$ by construction. Let us show that if $\mathrm{End}\, X_s$ does not embed into $M_g(\mathbb{Q})$, then there is <u>no</u> place $v \in \Sigma_f(\hat{K})$ for which $|\xi|_v < R_v(y_1,\ldots,y_{2g^2})$ (this implies globality). Indeed, if there were such a place v , then s and s_0 would have the same image in $\tilde{S}(\mathbb{F}_{p(v)})$ according to (3.1.1) (recall that $\tilde{S}$ denotes a regular projective model for S').
Since X/S descends to an Abelian scheme $\tilde{X}_{\tilde{U}/\tilde{U}}$ over some dense open subset $\tilde{U} \subset \tilde{S}$ (containing S), Gabber's lemma tells us that there exists a surjective projective morphism of integral normal schemes $\tilde{T} \xrightarrow{\psi} \tilde{S}$ such that the Abelian scheme $\psi^{-1}\tilde{X}_{\tilde{U}}$ on $\psi^{-1}\tilde{U}$ extends to a semi-Abelian scheme $\tilde{f} : \tilde{X} \longrightarrow \tilde{T}$ (i.e. a smooth separated morphism whose geometric fibers are extensions of Abelian varieties by tori), see P. Deligne "le lemme de Gabber" S.M.F. Astérisque 127 (1985) 131-150, in particular n°1.4, 1.7b and 4.10. Moreover this extension is unique because $\tilde{T}$ is normal. This implies: $\tilde{X} \times_{\tilde{T}} \psi^{-1}(S') = X' \times_{S'} \psi^{-1}(S')$. In particular,

$\tilde{X} \times_{\tilde{T}} \psi^{-1}(s)$ is the connected Néron model of X_s for any $s \in S$, and its special fiber at v is a torus whenever the images of s and s_0 in $\tilde{S}(\mathbb{F}_{p(v)})$ coincide. By general properties of connected Néron models,

$$(\mathrm{End}\, X_s) \otimes_{\mathbb{Z}} \mathbb{Q} = \mathrm{End}(\tilde{X} \times_{\tilde{T}} \psi^{-1}(s)) \otimes_{\mathbb{Z}} \mathbb{Q}$$

$$\longrightarrow \mathrm{End}(\tilde{X} \times_{\tilde{T}} \psi^{-1}(s)) \otimes_{\mathcal{O}_K} F_{p(v)}) \otimes_{\mathbb{Z}} \mathbb{Q} =$$

$$= \mathrm{End}(\mathbb{G}^g_{m|\mathbb{F}_{p(v)}}) \otimes_{\mathbb{Z}} \mathbb{Q} \simeq M_g(\mathbb{Q}) \; ;$$

there, the arrow is a priori only a ring (with unit) homomorphism of semi-simple $\mathbb{Q}$-algebras. But on using the Poincaré theorem on complete reducibility for Abelian varieties with multiplicative reduction, we may assume without loss of generality that X_s is simple (the Poincaré theorem is proved along the classical lines but using analytical methods of non-Archimedean uniformization, see L. Gerritzen "On non-Archimedean representation of Abelian varieties" Math. Ann. 196 (1972) 323-346). In this case, $(\mathrm{End}\ X_s) \otimes_{\mathbb{Z}} \mathbb{Q}$ is simple and the arrow is an embedding: this contradicts our assumption.

□

Now lemma 1.4 is just a combination of lemmata 3.3 and 3.4.

□

§ 4. SPECIAL POINTS ON SHIMURA VARIETIES AND OTHER COMMENTS

We shall interpret the theorem just proved in the language of Shimura varieties, and suggest a hierarchy of related problems which might possibly be attacked by G-function methods.

4.1 Throughout this paragraph, G denotes a connected linear algebraic group defined over $\mathbb{Q}$, whose radical is anisotropic over $\mathbb{R}$. Let H be a maximal compact algebraic subgroup of $G_{\mathbb{R}}$, and $D = H\backslash G_{\mathbb{R}}$ be the corresponding homogeneous space; there is a natural $G_{\mathbb{R}}$-invariant structure of C^{∞}-variety upon D (connected and simply-connected).
Let Γ be a neat arithmetic subgroup of G; because Γ is torsion-free, $\Gamma \cap xHx^{-1} = \{e\}$ for all $x \in G_{\mathbb{R}}$, thus Γ acts freely on D, whence a natural structure of C^{∞}-variety on the double coset space $H\backslash G_{\mathbb{R}}/\Gamma$.
We shall impose the following conditions:

H_1 : there exists a $G_{\mathbb{R}}$-invariant complex structure on D,

H_2 : the induced structure of complex manifold on $H\backslash G_{\mathbb{R}}/\Gamma$ is isomorphic to $Sh^{an}_{\mathbb{C}}$, for some quasi-projective algebraic variety Sh defined over $\bar{\mathbb{Q}}$ (which we call a "Shimura variety").
These conditions are automatically fulfilled when G is reductive (Baily-Borel-Faltings-Kazhdan), see [27]; in fact the $\bar{\mathbb{Q}}$-structure is uniquely determined if one imposes moreover that the special points are defined over $\bar{\mathbb{Q}}$ (the special points are the fixed points in D of maximal compact rational tori in G, or else

their images in D/Γ , which form a dense subset).
Let G' be an algebraic subgroup of G , also defined over $\mathbb{Q}$; let $g \in G_{\mathbb{R}}$ be such that $H' := G'_{\mathbb{R}} \cap g^{-1}Hg$ is a maximal compact subgroup of $G'_{\mathbb{R}}$. Then the map $H'x' \longmapsto Hgx'$ defines as embedding: $D' := H' \backslash G' \hookrightarrow D$. Now if $(D', \Gamma' := \Gamma \cap G')$ satisfies H_1, H_2 , this embedding gives rise to a map of algebraic varieties $D'/\Gamma' \longrightarrow D/\Gamma$. We call its image a "Shimura subvariety". Special points are particular cases. When G is reductive however, the essential examples are obtained for reductive G' .

4.2 Let us remark that in theorem 1.3, the non-existence of any ring embedding $\operatorname{End} X_s \hookrightarrow M_g(\mathbb{Q})$ is certainly fulfilled when X_s is of CM type, i.e. $\operatorname{End} X_s \otimes \mathbb{Q}$ contains a commutative sub-algebra of degree $2g$ over $\mathbb{Q}$.
Let $G = Sp_{2g}$ for odd g , so that any associated Shimura variety is the base of a "universal" Abelian scheme $X \longrightarrow Sh$. The fibers of CM type just correspond to special points on Sh (the compact rational torus G' is the special Mumford-Tate group of the fiber). Let S be as an affine curve on Sh such that:

i) S meets the zero-dimensional boundary component of the Satake compactification of Sh ,

ii) the generic fiber of $X_S \longrightarrow S$ is absolutely simple. Then the theorem admits the following

VARIANT. There are only finitely many special points of bounded residual degree lying on S .

4.3 In fact, one may think that for $g > 1$, the boundedness of the degree is unnecessary; this leads in a more general setting to the following problem (here the reader should remember the density of special points):

PROBLEM 1. Assume that G is reductive, and let S be an algebraic curve (not necessarily complete) on $H \backslash G_{\mathbb{R}}/\Gamma$. If infinitely many special points lie on S , is S necessarily a Shimura subvariety?

(R. Coleman has raised a similar problem where $G = Sp_{2g}$ and S

is replaced by the moduli space of curves of genus g .) This problem seems to be an analog of Raynaud's theorem, which tells that if infinitely many torsion points of an Abelian variety A lie on a curve S supported by A , then S is a translation of an Abelian subvariety [51].

4.4 One can go further with thinking that, except for modular families, there are only finitely many exceptional fibers in any algebraic one-parameter family of Abelian varieties with simple geometric generic fiber. In the general setting of Shimura varieties, this leads to:

PROBLEM 2. Hypotheses are as in Pb.1. If S meets infinitely many Shimura sub-varieties, is S itself a Shimura subvariety?

4.5 We do not assume any longer that the algebraic group G is reductive, and consider a slightly different problem. Instead of the special points, let us introduce the subset ${}_{H\cap G(\mathbb{Q})}\backslash G(\mathbb{Q})/\Gamma$ of ${}_{H}\backslash G_{\mathbb{R}}/\Gamma$, which is of course also a dense subset.

PROBLEM 3. Let S be an algebraic curve on ${}_{H}\backslash G_{\mathbb{R}}/\Gamma$. If $S \cap \left(H \cap G(\mathbb{Q})\backslash G(\mathbb{Q})/\Gamma\right)$ is an infinite set, is S a Shimura subvariety?

In the special case $G = Sp_{2g}$, this translates into a problem about isogeneous fibers in one-parameter families of Abelian varieties. Let us now look at the case when G is commutative. Because G is then (by assumption) anisotropic over $\mathbb{R}$, one has ${}_{H}\backslash G_{\mathbb{R}} \approx \mathbb{R}^n$, and hypothesis H_1 forces $n = 2g$, $g \in \mathbb{N}$. Thus Γ is a lattice in $\mathbb{R}^{2g}$ and hypothesis H_2 gives that ${}_{H}\backslash G_{\mathbb{R}}/\Gamma$ is an Abelian variety A of dimension g . One checks easily that Shimura sub-varieties are just the translations of Abelian subvarieties. At last, ${}_{H\cap G(\mathbb{Q})}\backslash G(\mathbb{Q})/\Gamma \approx \mathbb{Q}\Gamma/\Gamma$ may be identified with the torsion in A . Therefore Raynaud's theorem is just an affirmative answer to problem 3 for a commutative group G !

APPENDIX: A NEW PROOF OF THE TRANSCENDENCE OF π

The argument (based on global relations) given in this chapter failed for $g=1$, the elliptic case, because some factor π could not be eliminated in the relation (2.3.3). We now show that the same argument can be reversed to prove the transcendence of π , without using therefore any exponential or elliptic function. An effective transcendence measure could be given along the same line.

Let us consider the elliptic pencil of IX 1.2.3:

$$X_x : \qquad Y^2Z = 4X^3 + \frac{27}{x-1}XZ^2 + \frac{27}{x-1}Z^3$$

with functional invariant $J = \frac{1}{x}$. We keep the notations ω_1, ω_2 , γ_1, γ_2 of loc. cit.

Let x_n be an algebraic number such that X_{x_n} has complex multiplication by the order $\mathcal{O}_n = \mathbb{Z} + ni\,\mathbb{Z}$ of conductor n (the choice of -4 for the discriminant of $\mathcal{O}_1$ is not essential, but simplifies matters in the sequel because $x_1 = 1 \in \mathbb{Q}$).

Fix n , let $K \supset \mathbb{Q}(x_n, \sqrt{-2}, i)$, and set $e :=$ image of ni under $\mathcal{O}_n \xrightarrow{\sim} \mathrm{End}_K X_{x_n}$.

The decomposition of $H_{DR}(X_{x_n} {}_{|K})$ into e-eigenspaces is as follows: there exists $s_n \in K$ such that

$$(1) \quad \begin{cases} e^*\omega_1 = ni\,\omega_1 \\ e^*(\omega_2+s_n\omega_1) = -ni(\omega_2+s_n\omega_1) \end{cases} \quad \text{at } x = x_n .$$

This is "purely algebraic". But let us now choose a complex place v of K , corresponding to a complex imbedding $K \overset{i_v}{\hookrightarrow} \mathbb{C}$. After integration, (1) yields an element $\tau_v \in i_v(\mathbb{Q}(i))$ - depending on v - such that:

$$(2) \quad \begin{cases} \int_{\gamma_2} \omega_1 = \tau_v \int_{\gamma_1} \omega_1 \\ \int_{\gamma_2} (\omega_2+s_n\omega_i) = \bar{\tau}_v \int_{\gamma_2} (\omega_2+s_n\omega_1) \end{cases} \quad \text{at } x = i_v(x_n) .$$

Moreover the Wronskian relation

$$y_1 \frac{dy_2}{dx} - y_2 \frac{dy_1}{dx} = \frac{1}{144 \; i\pi x}$$

allows to transform (2) into

$$(3) \qquad \left[(\bar{\tau}_v - \tau_v) y_1 (\frac{dy_1}{dx} + i_v(s_n) y_1) - \frac{1}{144 \; i\pi x} \right] (i_v(x_n)) = 0 \; .$$

From the characterization of real singular moduli and the Fourier expansion of $J_{[44]}$, it follows that $i_v(J(X_{x_n})) \notin \,]-\infty,-1]$ if $n >> 0$. Hence $\tau_v = y_2/y_1(i_v(x_n))$ belongs to the fundamental domain of $SL_2(\mathbb{Z})$, after the remark at the end of 1.2.3; moreover τ_v is the unique point in this fundamental domain such that $X_{x_n} \times_{K,i_v} \mathbb{C} \simeq \mathbb{C}/_{\mathbb{Z}+\tau\mathbb{Z}}$. By the theory of modular transformations (together with the fact that $x_1 \in \mathbb{Q}$), one knows [44] that these τ_v - for various v - belong to the set $\left\{ ni, \frac{i+m}{n} \text{ (for } m = -\left[\frac{n}{2}\right], \ldots, n-1-\left[\frac{n}{2}\right]) \right\}$. The <u>key point</u> is that $\bar{\tau}_v - \tau_v$ takes only <u>two</u> values, namely $\frac{2n}{i}$ and $\frac{2}{ni}$. Now assume that π were algebraic, with conjugates $\pi_1, \ldots, \pi_\delta$, and extend K so that all $\pi_j \in K$. I claim that

$$(4) \quad \left[\prod_j \left(2ny_1 (\frac{dy_1}{dx} + s_n y_1) - \frac{1}{144\pi_j x_n} \right) \left(\frac{2}{n} y_1 (\frac{dy_1}{dx} + s_n y_1) - \frac{1}{144\pi_j x_n} \right) \right] (x_n) = 0$$

is a <u>global relation</u> at x_n, of degree 4δ, between the G-functions y_1 and dy_1/dx.[(1)]
Indeed it is satisfied at every $v \in \Sigma_\infty(K)$ because of (3), and there is <u>no</u> finite place to consider because

$$|x_n|_v = \left| \frac{1728}{j(X_{x_n})} \right|_v \geq |1728|_v \text{ (since } X_{x_n} \text{ has complex multiplication),}$$

and one checks that $|1728|_v = R_v(y_1)$!
By the Hasse principle for values of G-functions, we conclude that $h(x_n)$ is bounded independently of n.
But this <u>contradicts</u> a result of P. Cohen which tells that for n prime, $h(x_n) \sim 6 \log n$, ("On the coefficients of the transformation

(1) algebraically independent, by ordinary differential Galois theory.

polynomials for the elliptic modular function", Math. Proc. of Cambridge Phil. Soc. 95 (1984) 389-402, – this can be also proved using Faltings' height $\left(X_{x_n} \right) \sim \frac{1}{12} h(x_n)$, and the isogeny formula).

Hence one cannot square the circle.

□

Bibliography

Caution: This is by no means an exhaustive list of papers about G-functions; further references to the extensive literature on this topic can be found in [7], [18], or [60].

[1] Y. André, "Multiplication complexe dans les pinceaux de variétés abéliennes" Séminaire de théorie des nombres de Paris 84/85, Birkäuser Boston, Progress in Math. 63 (1986) 1-22. Erratum in vol 71(1987) 209.

[2] Y. André, "Quatre descriptions des groupes de Galois différentiels", (1987), in the proceedings of the séminaire d'algèbre de Paris 86/87" Springer L.N. 1296.

[3] Y. André, "Mumford-Tate groups and the theorem of the fixed part", preprint Max-Planck-Institut, Bonn.

[4] Bateman Manuscript Project, Higher transcendental functions, Mc Graw Hill Co, New York 1953.

[5] P. Berthelot, A. Ogus, "F-isocrystals and De Rham cohomology I" Inv. Math. 72 (1983) 159-199.

[6] D. Bertrand, F. Beukers, "Equations différentielles linéaires et majorations de multiplicités" Ann. Scient. Ec. Norm. Sup. 4 Série, t 18 (1985) 181-192.

[7] E. Bombieri, "On G-functions", Recent progress in analytic number theory, Academic Press (1981), vol 2, 1-67.

[8] E. Bombieri, A. Sperber, "On the p-adic analyticity of solutions of linear differential equations" Illinois J. of Math. vol 26 n° 1 (1982) 10-18.

[9] R.H. Calderon, W.T. Martin, "Analytic continuations of diagonals and Hadamard composition of multiple power series", Trans. Amer. Math. Soc. 44 (1938) 1-7.

[10] G. Christol, "Sur une opération analogue à l'opération de Cartier en caractéristique nulle" C.R.A.S. 271 (1978) Série A, 1-3.

[11] G. Christol, "Structure de Frobenius faible des équations différentielles p-adiques", Groupe d'étude d'analyse ultramétrique 1975/76, Secrétariat mathématique, Institut H. Poincaré, Paris.

[12] G. Christol, "Systémes différentiels linéaires p-adiques: structure de Frobenius faible" Bull. Soc. Math. France 109 (1981) 83-122.

[13] G. Christol, "Un théorème de transfert pour les disques singuliers réguliers", S.M.F. Astérisque 119/120 (1984) 151-168.

[14] G. Christol, "Diagonales de fractions rationnelles et équations de Picard-Fuchs" Groupe d'étude d'analyse ultramétrique 1984/85 Secrétariat mathématique, Institut H. Poincaré, Paris.

[15] G. Christol, "Fonctions hypergéométriques bornées", Groupe d'étude d'analyse ultramétrique 1986/87 Secrétariat mathématique, Institut H. Poincaré, Paris.

[16] G. Christol, "Diagonales de fractions rationnelles" Preprint Paris 1987.

[17] G.V. Chudnovsky, "Rational and Padé approximations to solutions of linear differential equations and the monodromy theory", Complex analysis, microlocal calculus and relativistic quantum theory, Les Houches (1979) Berlin, Heidelberg, New York, Springer Verlag 1980 (L.N. in physics 126), 126-169.

[17 1/2] G.V. Chudnovsky, "Algebraic independence of values of exponential and elliptic functions", Proc. ICM, Helsinki (1978).

[18] D.V. Chudnovsky, G.V. Chudnovsky, "Applications of Padé approximations to disphantine inequalities in values of G-functions" Lect. Notes in Math. 1052, Berlin, Heidelberg, New York, Springer-Verlag (1985) 1-51.

[19] D.V. Chudnovsky, G.V. Chudnovsky, "Applications of Padé approximations to the Grothendieck conjecture on linear differential equations", Lect. Notes in Math. 1135, Berlin Heidelberg, New York, Springer-Verlag (1985) 52-100.

[19 1/2] D.V. Chudnovsky. G.V. Chudnovsky. "Padé approximations and diophantine geometry", Proc Nat. Ac. Sc. USA 82, (1985), 2212-2216.

[20] P. Debes, "G-fonctions et théoréme d'irréductibilité de Hilbert" Acta Arithm. 47 (1986) 371-402.

[20 1/2] P. Debes, "Résultats récents liés au théorème d'irréductibilité de Hilbert" Séminaire de théorie des nombres de Paris 1985/86, Birkhäuser, Boston, progress in Math. 71 (1987) 19-37.

[21] P. Deligne, Equations différentielles à points singuliers réguliers, Lect. Notes in Math. 163, Berlin, Heidelberg, New York, Springer-Verlag (1970).

[22] P. Deligne, "Théorie de Hodge", II: Publ. Math. IHES 40 (1972) 5-57; III: Publ. Math. IHES 44 (1974) 5-78.

[23] P. Deligne, and al., Hodge cycles, Motives and Shimura Varieties; I: P. Deligne "Hodge cycles on Abelian varieties" 9-100 (notes by J. Milne); II: P. Deligne and J. Milne, "Tannakian categories" 101-228. Berlin, Heidelberg, New York, Springer-Verlag, Lect. Notes in Math.900 (1982).

[23 1/2] P. Deligne, "Variëtés de Shimura: interprétation modulaire, et techniques de construction de modèles canoniques", Automorphic forms, Representations, and L-functions; Proc. symp. Pure Math., vol 33, part 2, AMS, Providence R.I. 1979, 247-289.

[24] P. Deligne, "Théorême de Lefschetz et critères de dégénérescence de suites spectrales" Publ. Math. IHES 35, (1968) 107-126.

[24 1/2] P. Deligne, L. Illusie, "relèvements modulo p^2 et décomposition du complexe de De Rham" Inv. Math. 89 (1987) 247-270.

[25] B. Dwork, P. Robba, "Effective p-adic bounds for solutions of homogeneous linear differential equations". Trans. Am. Math. Soc. 259 (1980) 559-577.

[26] A. Escassut, P. Robba, Algèbres de Banach ultramétriques et prolongement analytique, SMF Astérisque 10 (1973).

[27] G. Faltings, "Arithmetic varieties and rigidity" in Sém. de théorie des nombres de Paris (1982/83), Birkhäuser Boston, 63-77.

[28] M. Fliess, "Sur divers produits de séries formelles", Bull. Soc, Math. Fr. 102 (1974) 181-191.

[29] H. Furstenberg, "Algebraic functions over finite fields" J. of Alg 7 (1967) 271-277.

[30] A.I. Galočkin, "Estimates from below of polynomials in the values of analytic functions of a certain class" Math. USSR Sbornik 24 (1974) 385-407. Original article in Mat. Sbornik 95 (137) (1974).

[31] F. Gantmacher, Théorie des matrices 2. Dunod, Paris (1966).

[32] A. Grothendieck, "On the De Rham cohomology of algebraic varieties", Publ. Math. IHES 29 (1966) 95-103.

[33] R. Hartshorne, "On the De Rham cohomology of algebraic varieties", Publ. Math. IHES 45 (1976) 5-99.

[34] R. Hartshorne, Algebraic Geometry, Berlin, Heidelberg, New York, Springer-Verlag (1977).

[35] L. Illusie, "Report on crystalline cohomology" Proc. Symp. Pure Math., vol 29, A.M.S. Arcata (1974) 459-478.

[36] N. Jacobson, Basic Algebra II, Freeman, San Francisco (1980).

[37] N. Katz, "Travaux de Dwork" Sêminaire Bourbaki, 1971/72, exp 402, 167-200. Springer-Verlag, Lect. Notes in Math. 317 (1973).

[38] N. Katz, "On differential equations satisfied by period matrices", Publ. Math. IHES 35 (1968).

[39] N. Katz, "Nilpotent connections and the monodromy theorem: application of a result of Turritin". Publ. Math. IHES 39 (1970) 175-232.

[40] N. Katz, "Algebraic solutions to differential equations" Inv. Math. 18 (1972) 1-118.

[41] N. Koblitz, p-adic numbers, p-adic analysis and Zeta functions. Springer-Verlag GTM 58 (1977).

[42] S. Lang, Introduction to transcendental numbers, Addison-Wesley, Reading, London, Sydney (1966).

[43] S. Lang, Algebraic number theory, Addison-Wesley, Reading, London, Sydney (1970).

[44] S. Lang, Elliptic functions, Addison-Wesley (1973).

[45] S. Lang, Fundamentals of diophantine geometry, Springer-Verlag (1983).

[46] M. Laurent, "Une nouvelle dêmonstration du thêorème d'isogênie, d'après D. et G. Chudnovsky", Sêminaire de Thêorie des Nombres,Paris 1985/86, 119-129, Birkhäuser.

[47] P. Landweber, "Elliptic cohomology and modular forms", Preprint 1986, Rutgers University.

[48] Y. Manin, "Moduli Fuchsiani", Annali scuola Norm. Sup. Pisa, sur III 19 (1965) 113-126.

[49] P. Monsky, "Finiteness of De Rham cohomology", Am. J. of Math. 94 (1972) 237-245.

[50] G. Mustafin, "Families of algebraic varieties and invariant cycles" Math. Izv. 27 n°2 (1986); original paper (russian) in Izv. Acad. Nauk. 49 (1985) n°5.

[51] M. Raynaud, "Courbes sur les variétés abéliennes", Inv. Math. 71 (1983) 207-233.

[52] C. Runge, "Über ganzzahlige Lösungen von Gleichungen zwischen zwei Veränderlichen", J. Reine Angew. Math. 100 (1887) 425-435.

[53] W. Schmidt, "Variation of Hodge structure: on the singularities of the period mapping", Inv. Math. 22 (1973) 211-319.

[54] J.P. Serre, Corps locaux, Hermann, Paris 1962.

[54 1/3] J.P. Serre, Géométrie algébrique et géométrie analytique" (GAGA), Ann. Inst.Fourier, Grenoble 6 (1956).

[54 2/3] J.P. Serre, "Quelques propriétés des groupes algébriques commutatifs", Astérisque 69-70 (1979).

[55] A. Shidlovsky, "Diophantine approximations and transcendental numbers", (in Russian), Publ. Univ. Moscow (1982).

[56] C.L. Siegel, "Über einige Anwendungen diophantischer Approximationen". Abhandlungen der Preussischen Akademie der Wissenschaften. Phys. Math. Kasse 1929 nr.1. Also "Gesammelte Abhandlungen I", Springer-Verlag (1966) 209-266.

[57] V.G. Sprindzuk, "Arithmetic specializations in polynomials" J. Reine Angew. Math. 340 (1983) 26-52.

[58] J. Steenbrink, "Limits of Hodge structures", Inv. Math. 31 (1976) 229-257.

[59] P.F. Stiller, Automorphic forms and the Picard number of an elliptic surface", Aspects of Math. E5, Vieweg, Wiesbaden (1984).

[60] K. Väänänen, "On a class of G-functions", Séminaire de théorie des nombres de Paris 1981/82, Birkhäuser Boston, Progress in Math. 38, 313-319.

[61] G. Wüstholz, "Algebraic groups, Hodge theory and transcendence" Preprint Max-Planck-Institut, Bonn (1986).

[62] S. Zucker, "Degeneration of Hodge bundles, after Steenbrink" Topics in transc. alg. Geometry, P. Griffiths ed., Annals of Math. Studies, Princeton.

[63] "Padé approximants and its Applications", Amsterdam 1980 Proceedings edited by M.G. de Bruin and H. van Rossum, Lect. Notes in Math. 888, Springer-Verlag.

Index

Glossary of Notations

See also the "general notations" at the beginning of the book.

$(a)_n = a(a+1) \dots (a+n-1)$	
$A_1(k) = k[x, d/dx]$	the Weyl algebra
$A_v(a,r)$	the ring of v-adic analytic functions in the disk $D(a,r)$
α, β	order of elements of $k((x))$ at 0
C, x^C	"exponential" matrix with constant entries
d, d_v	degree of a number field; local degree
$\partial = x\, d/dx$	
Δ_ν	diagonalization map
E, E_0 ...	algebras of analytic elements
F, G, Γ	matrices associated with differential operators on ∂-modules
h, h_f, h_∞	the invariant logarithmic height; its finite (resp. infinite) part
Hg^n	Hodge rings
$i, j, \dots, n$	running indices
k	a field of characteristic 0, often a complete valued one
K	a number field
L_k	polylogarithms
L, Λ	linear differential operators
$\mathcal{M}$, $\mathcal{N}$	∂-modules
μ	order of a differential equation, rank of a ∂-module
$\odot n$	n^{th} symmetric power
N	a large parameter; or a nilpotent matrix, e.g. $\frac{1}{2i\pi}$ log (local monodromy)
Φ, Ψ, Ξ	Frobenius' functor, its inverse, and Christol's functor

$\prod$, $\circledast$	Fonctors "Isom", see IX 2.3 amd 3.1
$P_X^{\cdot}$	the comparison isomorphism between algebraic De Rham- and rational cohomology (see IX)
q, u	polynomials
Rf_*^{DR}	higher direct image in De Rham cohomology (see II)
R_v	v-adic radius of convergence (resp. solvability) of a series (resp. ∂-module)
ρ	global radius of series or ∂-modules
σ	size of series or ∂-modules
Sin Λ	set of finite non-apparent singularities of a differential operator Λ ; in general, we set
$s = \lvert \mathrm{Sin}\ \Lambda \rvert$	
t_v	generic point over $\mathbb{C}_v$
τ	a small parameter
θ	solution of a ∂-module
v,w	places of a number field
$\underline{x} = (x_1,\ldots,x_\nu)$	ν-tuple of independent variables
$X = Yx^C$	solution of a differential system at a logarithmic singularity
y	often, a G-function
Y	sometimes, an element of $K[[x]]^\mu$
ζ,ξ	algebraic numbers
$\lvert\ \rvert$ or $\lvert\ \rvert_v$	a v-adic absolute value (= multiplicative norm), e.g. Gauss' absolute value; for $v\mid\infty$, $\lvert\ \rvert$ denotes the Euclidean absolute value
$\lVert\ \rVert_v$	v-adic norm on matrix spaces
$\frac{1}{(2i\pi)^k}\int_Y \omega$	periods of a smooth projective variety
*	as an exponent: indicates a duality; elsewhere, Hadamard product

Aspects of Mathematics

English-language subseries (E)

Vol. E1: G. Hector/U. Hirsch, Introduction to the Geometry of Foliations, Part A

Vol. E2: M. Knebusch/M. Kolster, Wittrings

Vol. E3: G. Hector/U. Hirsch, Introduction to the Geometry of Foliations, Part B

Vol. E4: M. Laska, Elliptic Curves over Number Fields with Prescribed Reduction Type

Vol. E5: P. Stiller, Automorphic Forms and the Picard Number of an Elliptic Surface

Vol. E6: G. Faltings/G. Wüstholz et al., Rational Points
(A Publication of the Max-Planck-Institut für Mathematik, Bonn)

Vol. E7: W. Stoll, Value Distribution Theory for Meromorphic Maps

Vol. E8: W. von Wahl, The Equations of Navier-Stokes and Abstract Parabolic Equations

Vol. E9: A. Howard/P.-M. Wong (Eds.), Contributions to Several Complex Variables

Vol. E10: A. J. Tromba, Seminar on New Results in Nonlinear Partial Differential Equations
(A Publication of the Max-Planck-Institut für Mathematik, Bonn)

Vol. E11: M. Yoshida, Fuchsian Differential Equations
(A Publication of the Max-Planck-Institut für Mathematik, Bonn)

Vol. E12: R. Kulkarni, U. Pinkall (Eds.), Conformal Geometry
(A Publication of the Max-Planck-Institut für Mathematik, Bonn)

Vol. E13: Y. André, G-Functions and Geometry
(A Publication of the Max-Planck-Institut für Mathematik, Bonn)

Vol. E14: U. Cegrell, Capacities in Complex Analysis

Aspekte der Mathematik

Deutschsprachige Unterreihe (D)

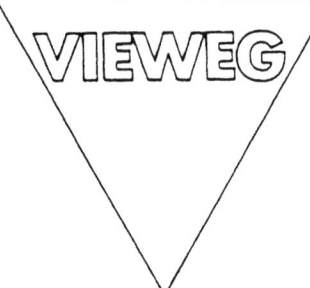

MIX
Papier aus verantwortungsvollen Quellen
Paper from responsible sources
FSC® C105338

If you have any concerns about our products,
you can contact us on
ProductSafety@springernature.com

In case Publisher is established outside the EU,
the EU authorized representative is:
Springer Nature Customer Service Center GmbH
Europaplatz 3, 69115 Heidelberg, Germany

Printed by Libri Plureos GmbH
in Hamburg, Germany